Reviving the Dying Giant

Shahrooz Mohajeri • Lena Horlemann
Editors

Reviving the Dying Giant

Integrated Water Resource Management in the Zayandeh Rud Catchment, Iran

Springer

Editors
Shahrooz Mohajeri
inter 3 GmbH Institute for
Resource Management
Berlin, Germany

Lena Horlemann
inter 3 GmbH Institute for
Resource Management
Berlin, Germany

ISBN 978-3-319-54920-0 ISBN 978-3-319-54922-4 (eBook)
DOI 10.1007/978-3-319-54922-4

Library of Congress Control Number: 2017941084

Printed on acid-free paper

This Springer imprint is published by Springer Nature
The registered company is Springer International Publishing AG
The registered company address is: Gewerbestrasse 11, 6330 Cham, Switzerland

Preface

In Iran's 6th Five-Year Development Plan, water resources are defined as one of three critical issues for the country's future development. Water is a vital resource for the country's overall development and for maintaining public health, but its availability has been put at risk by recent developments.

Iran is located in an arid and semiarid area which represents a natural restriction of water resources availability. In addition, the country has faced 15 consecutive years of droughts which have intensified water scarcity. In this time, renewable water resources have declined from 130 to 105 billion m^3. Climate change has also significantly changed the precipitation patterns: Instead of a few long periods of rainfall, today there are several short periods. Combined with rising temperatures this means that a significant amount of rain water evaporates before reaching the aquifers. Despite these issues, water scarcity has been aggravated by a dramatic increase in water demand for agricultural, domestic, and industrial purposes.

These developments have revealed that the water availability for people and the environment cannot be addressed by technological solutions only but requires a holistic management approach. Thus, in the new government, a paradigm shift in water policies has taken place and the Ministry of Energy is no longer promoting hardware approaches like dam construction. Instead, a number of priorities have been defined which include water consumption management, the establishment of intelligent networks, pollution prevention, the optimization of wastewater treatment plants, or water recycling. Along with the technical solutions, integrated approaches, participation of consumers in water management, and the establishment of coordination councils (RBOs) are expected to have positive effects on Iran's water resources.

In the history of the country's water and wastewater industry, German companies, as the main foreign partners, have always been with us, and cooperation has been very successful in various sectors. The new water policy priorities have set the grounds for further Iranian-German cooperation. This cooperation does not only include transfer of technologies, but also investment and an exchange of know-how and management concepts. Undoubtedly, the new political conditions provide an excellent basis for extensive cooperation with German companies and organizations.

The Iranian-German cooperation in the "IWRM Zayandeh Rud" project reflects the paradigm shift in Iran's water management as it combines technological and nontechnological approaches. The German Ministry of Education and Research has been funding this project since 2010 and we would like to express our gratitude for this effort. As the highest authority of water resources in the country, we can assure full support of the government for further Iranian-German endeavors in the field of water management.

Minister of Energy Iran
Tehran, Iran

Hamid Chitchian

Acknowledgments

We would like to express our gratitude to the German Federal Ministry of Education and Research (BMBF) for funding this project and thereby enabling this fruitful German-Iranian cooperation. Further thanks go to the Project Management Agency Karlsruhe for its support during the project realization. We also like to thank the Iranian Ministry of Energy and the Iranian Water Resources Management Company as well as the National Water and Wastewater Engineering Company. It is to their credit that the IWRM Zayandeh Rud project has attracted more and more attention over the years. Moreover, we like to thank the Isfahan Governor's Office, the Water and Wastewater Companies in Isfahan and Kashan, the Jihad Agricultural Organization of Isfahan, the Isfahan Environmental Protection Organization, the Regional Water Company Chaharmahal-va-Bakhtiari, the Mirab Zayandeh Rud Company, the Isfahan Higher Education and Research Institute, Zayandab Consulting Engineers, the Zayandeh Rud Urban Development Organization, the Industrial Settlement Organization, and the Isfahan University of Technology as well as the large industries in Isfahan Province who have all made this project possible.

Special thanks go to the Isfahan Regional Water Company, particularly Mr. Torfeh and Dr. Mirmohammad Sadeghi who has taken over the reins. We would like to thank the IWRM project management at the Isfahan Regional Water Board Company, our current project manager Mr. Heydarpour and Mr. Asady, who was project manager from 2010 to 2015, and their entire team.

We are also grateful to the IWRM commission and all its former and current members who have accompanied the project with their valuable and indispensable expertise. We would like to thank Mr. Aghili who supported the project team by translating important documents, interpreting in meetings, and accompanying German partners during field trips.

Last but not least, the editors thank Springer Nature, Dr. Stephan Klapp and Judith Terpos in particular, for their support in publishing this book.

Contents

Chapter 1
Introduction

Lena Horlemann and Shahrooz Mohajeri

All work is a seed sown; it grows and spreads, and sows itself anew. Persian proverb

At the World Summit on Sustainable Development in Rio de Janeiro 1992 the concept of IWRM was established as the international guiding principle for managing increasingly scarce water and its related resources. IWRM highlights the coordinated development and management of water at the watershed level (GWP 2000) and aims at ecological, sectoral as well as regulatory integration. It ideally replaces command-and-control approaches and promotes a shift toward participatory, bottom-up decision-making procedures. Agenda 21 and the Dublin Principles form its implementation framework. In response to failed sectoral approaches, IWRM emblematizes a paradigm shift to integrated and basin wide thinking: Water bodies and their related ecosystems, water using sectors and different governance levels have to be considered in management decisions. Ecological, economic and social objectives must be linked together in order to achieve sustainable outcomes. The World Bank summarizes:

> Today, it is considered best practice in water resources planning to integrate water quantity and quality management for both groundwater and surface water, while incorporating a full understanding of how the natural resources and the people of a basin are impacted by various levels of development or by adopting new resource use policies. Land use as well as land and vegetation management are thus issues that need to be considered in water resources planning and management. (The World Bank 2006, p. 3).

IWRM requires a meaningful mixture of management approaches for the respective water basin, ranging from participation of public and private stakeholders, the creation of adequate institutional structures to the implementation of new technologies (GWP 2003; The World Bank 2006; GWP 2009; UNEP 2012). The challenges of defining this "meaningful mixture" in general and particularly for

L. Horlemann (✉) • S. Mohajeri
inter 3 GmbH Institute for Resource Management, Berlin, Germany
e-mail: horlemann@inter3.de

S. Mohajeri, L. Horlemann (eds.), *Reviving the Dying Giant*,
DOI 10.1007/978-3-319-54922-4_1

specific regions or river basins have been intensely described and discussed (see for example Mostert 2003; Biswas 2004; Kalbus et al. 2012; Hering and Ingold 2012; Martinez-Santos et al. 2014; Borchardt et al. 2016).

1.1 Research on Behalf of the German Government

The aim of the Federal Ministry of Education and Research (BMBF) is to promote the innovative potential of German science and technology to develop solutions for water availability and solving water quality problems.

The main aims are:

- Better access to clean water and sanitation,
- Higher water productivity and more efficient water use,
- Improved water quality.

To achieve these goals, BMBF launched the funding priority "Integrated Water Resources Management (IWRM)".[1] In 16 research projects in Europe, the Middle East, Asia, Africa and Latin America various approaches, methods and technologies have been looked at to discern what is necessary and feasible for IWRM and how management concepts can be adjusted to local conditions. By developing integrated planning tools, adjusting water technologies to different climatic, economic and social conditions and through multifarious training activities, the BMBF contributes to the United Nation's Sustainable Development Goals (SDG). As a Middle Eastern country with severe water challenges, Iran is one of the focal areas of BMBF research activities where the implementation of IWRM can produce valuable results for the entire region.

1.2 Water Management Challenges in Iran

Despite general progress in Iran's water sector the situation in most parts of the country is still far from ideal. In the face of dynamic demographic development, rising per-capita water consumption, a lack of funds and skilled personnel, the challenges are vast. Due to increasing water scarcity, enormous efforts are needed to improve the tense ecological and social situation. Political reforms and technological modernization are necessary in order to realize efficient water use and monitoring of water quality and quantity.

Drinking water demand has grown continuously. Recently, water consumption has increased more than population growth. While between 1995 and 2000 the population grew by 13.3% and therefore twice as much as water consumption, the

[1] see http://www.bmbf.wasserressourcen-management.de/en/index.php.

situation reversed between 2000 and 2005: Drinking water demand increased by 30% while the population growth decelerated to 9.5% during these 5 years.

The growing need for action stands vis-à-vis financially and operationally weak water sector institutions. Due to low fees that are fixed by the central government, water and wastewater companies face increasing income losses. Between 1994 and 2008, the losses of all 35 urban water and wastewater companies accumulated to 1.5 billion US$. A lack of regional decision-making power and independent supervision adds to the climate of insecurity and vague responsibilities. On paper, urban water and wastewater companies are organized under private law and are largely autonomous. In practice, however, the government and provincial or urban administrations, as the main shareholders, influence important decisions like human resources management or investments.

Another obstacle for the effective development of the Iranian water sector is the lack of skilled personnel. Particularly in cities and regions far away from the booming regions of Tehran, Isfahan or Western Iran there are not enough experienced engineers and skilled workers for building and operating technologically sophisticated sites and plants. This situation is particularly serious as the various regions in Iran show different hydrogeological conditions which require specific and adjusted rather than standardized solutions. There is a general lack of practice-oriented training to develop adequate numbers of skilled personnel with appropriate qualifications who would run such customized plants. Yet, where there is no experience and know-how about innovative technologies and processes, there is no meaningful investment. Moreover, the durability of plants that are not properly maintained is significantly shortened.

The water management deficits of the last decades have led to social unrest in recent years which the government has taken seriously. Some countermeasures have been introduced, but often not in the desired timeframe or without sustainable comprehension, like water transfers or compensation for yield losses.

1.3 The Zayandeh Rud: The Lifeblood of an Entire Region

The catchment of the Zayandeh Rud (Farsi for 'life-giving river'), is one of the most diverse regions in Iran: From the snow-covered Zāgros Mountains, through wide floodplains and desert regions to the Gavkhuni wetland, the catchment is a special habitat. For centuries the river has attracted people and the entire region is host to a uniquely diverse ecology.

During the last 60 years, the population in the catchment has grown from less than a million to more than four million. Today, more than 1 million people live from the land, producing wheat and other staple food. Important steel, oil and cement industries have settled along the river which along with numerous smaller enterprises employ more than 300,000 people.

The steady growth of the region, coupled with the onset of climate change, have taken their respective tolls, leading to increasing water management challenges.

While water demand rises, the Zayandeh Rud's water resources decrease and with them the livelihood of people and important ecosystems dwindle. Temperatures have been rising constantly, while annual rainfall has been declining.

Up until a few years ago the river dominated the cityscape of Isfahan. Its historical bridges and little canals were famous tourist attractions and places of recreation for young and old alike. Numerous species of birds migrated there for the winter in the region around the Gavkhuni wetland and even ventured into the city centre. For flamingos the salty lake was an ideal habitat.

As the gap between water availability and water demand grows the different water users increasingly compete for the scarce resource. Farmers, subject to uncertainty, ask themselves, "When will the next drought come? Will I have enough water for my fields? Will I be able to feed my family?" Industrial and drinking water supply is also under threat. Up until today, industrial enterprises in the catchment are still dependent on fresh water resources for various purposes, and their demand is rising with steady industrial growth. Last but not least, the Zayandeh Rud is the main supplier for drinking water in the region, also for cities in provinces other than Isfahan, like Yazd.

The diminishing of 'the life-giving river' is a tragedy for the entire region and not only for economic reasons but also from social, health and ecological perspectives. This situation that has evolved over years, however, will not be solved overnight. But it is worthwhile being tenacious in alluring all stakeholders for sustainable water management.

A sustainable and integrated water management concept for the Zayandeh Rud catchment is therefore fundamental. The natural beauty of the region, its biodiversity and not to mention more than 4.3 million people will benefit collectively from efficacious solutions. In this regard the Zayandeh Rud is representative of the plight of humans and their environment in many parts of Iran and in the Middle East in that it is dominated by a dry climate and water scarcity.

1.4 The Research and Development Project "Integrated Water Resources Management Zayandeh Rud"

Within the "IWRM Zayandeh Rud" project Iranian and German partners have been developing and implementing a generally accepted water management concept for the entire catchment. The German consortium involves seven scientific institutes and companies specialised in the water sector. In Iran the Isfahan Regional Water Company has been coordinating the project including all relevant stakeholders on national, regional and sectoral levels. Both partner countries combine their skills to make water use along the river sustainable and to balance the competing interests of the different water users.

The aim of the project is to present options for integrated water management in Iran and the entire region using the example of the Zayandeh Rud catchment. The project team has been working towards this goal since 2010.

A necessary precondition for IWRM is that the relevant actors of the affected provinces, sectors and national institutions become aware that they do not only sit in one boat but also have to row in the same direction. IWRM requires discussion and agreement on the measures to be taken and to cooperate in order to reach a common goal. Therefore, a project structure was developed that involved all relevant stakeholders and water using sectors.

The first step of the project included understanding the overall structure in Iran's water sector including institutions, stakeholders and practices (see Chap. 2). Also, the project partners had to get a picture of the specific importance of the Zayandeh Rud catchment as the focus of research (see Chap. 3). It quickly became obvious that there was not only a large number of stakeholders that had to be taken into account in an IWRM concept, but that there were also many conflicting opinions about the most pressing challenges in the Zayandeh Rud catchment. A mutual problem definition was necessary – and could be achieved in a participatory process (see Chap. 4). It was also important to gather information about the existing capacities that are essential for establishing sustainable water management, hence another research activity focused on the field of training in Iran's water and wastewater sector (see Chap. 5).

But how can the complexity of the situation in the Zayandeh Rud catchment be adequately understood? What guarantee is there that the solution to one problem will not just lead to another problem? And how can we make sure that decisions that seem adequate from a water management perspective become acceptable from political and social perspectives as well?

Decisions which are both reasonable with regards to water management as well as being politically and socially accepted, need a solid scientific basis. As the basis for the IWRM process the project team developed an IWRM concept for the Zayandeh Rud catchment (see Fig. 1.1) together with, and accepted by, the main stakeholders. The German and Iranian project partners also agreed upon developing an IT based decision support system (DSS) based on a Water Management Tool (WMT, see Chap. 12).

The IWRM concept provides a step-by-step process where the main national, regional and local stakeholders identify the problems in the catchment and collect ideas for their solutions.

This is how it works:

- An idea, for example the withdrawal of water access rights, is fed into the DSS. Now the DSS can show how this proposed measure will impact upon water resources in the catchment.
- Based on this, specific recommendations can be developed which should be adjusted to a vision or overall strategy for the catchment. Eventually a water management decision should be taken which is acceptable for ministers and farmers alike.

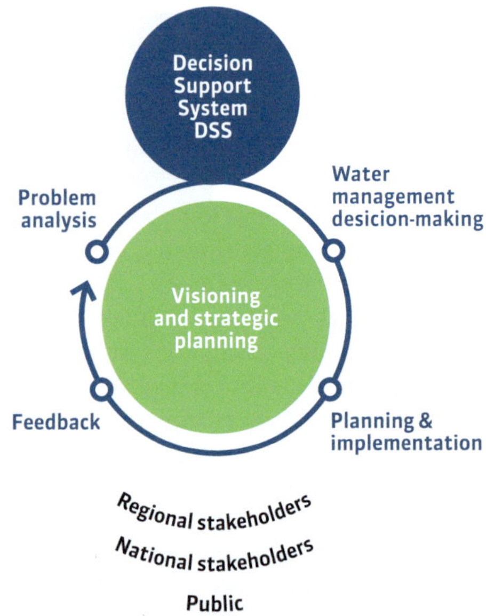

Fig. 1.1 Adapted IWRM concept for the Zayandeh Rud catchment (böing-gestaltung/inter 3)

- Built on a good and acceptable decision, the planning and implementation of specific measures can be carried out and their status and impact need to be monitored continuously.
- Depending on how well a measure has taken effect, it can be adjusted and ideas on problem solving can be redefined.

In the same way the stakeholders were able to agree upon their expectations of a Water Management Tool for the catchment. This system is able to process large amounts of data and provide information about the possible results of management decisions through models. It is neutral in its analysis, clear in the presentation of its results and therefore the ideal tool for steering an IWRM process.

In order to understand the situation of water resources and to be able to work with information in a target-oriented way, current and future water availability and use of the main water using sectors need to be known. At the beginning of the project, however, there were no coherent data available for the Zayandeh Rud catchment (e.g. on water consumption of the single sectors) which were clear and acceptable for the different stakeholders. Thus, certain decisions always remained a source of conflict. In close cooperation with the Iranian partners, data were collected and harmonized for the agricultural sector (see Chap. 6), the industrial sector (see Chap. 7), and the urban water sector (see Chap. 8). Moreover, water quality and existing monitoring techniques were examined (see Chap. 9). Research activities also included information on the current state of the wastewater sector (see Chap. 10) and the vulnerability of farmers (see Chap. 11). In doing so, interrelations of water management challenges, management effects on water users (including the environment), current management approaches and feasible future options were assessed.

For gathering the necessary data to set up the Water Management Tool (see Chap. 12), the reference year 2006 (Persian year 1385) was used due to the fact that this was a relatively "normal" year with sufficient water available for all water users. In order to clarify the main influences on water resources and their interactions it was decided to develop four independent models and integrate them: a climate change model (see Chap. 13), a hydrological model (SWAT, see Chap. 14), a groundwater model (FEFLOW, see Chap. 15) and the MIKE Basin model which can depict changes in water availability in the catchment subsequent to specific management decisions (see Chap. 16).

1.5 Looking Ahead

The water management challenges in the Zayandeh Rud catchment are only one example of the same or similar problems in the entire country. The IWRM project is therefore also an example of approaches, tools and methods that can be used to implement IWRM in Iran. Necessary decisions and actions have to be taken by the Iranian decision-makers. German partners can only show possible ways to a locally adjusted IWRM based on their long standing experiences in Germany or other international projects.

It is obvious that the concepts and tools developed within the project are necessary for sustainable planning and use of water resources, but never enough. They can only come into effect if the decision-makers lead the way on a joint journey towards IWRM instead of achieving the maximum outcome for their own region or sector. The German partners will do their utmost to support this process. Precisely because all decisions in the region have to be taken by the Iranian partners, the unbiased and internationally experienced German team can moderate and mediate between the parties.

Molle et al. (2009, p. 207) underline that "some sort of basin-level coordination body is needed to analyse hydrological data, establish transparent allocation schemes (. . .), discuss priorities and development plans, and integrate representatives from the different socio-economic sectors." The establishment of the first River Basin Organization for the Zayandeh Rud in 2014 is without doubt an important step towards common thinking and action. Now it is time for developing a joint vision, defining the necessary measures for achieving the goals, implementing the measures step by step and evaluating their impact. Molle et al. (ibid.) add, however, that "establishing a sound water regime at the basin level is thus a monumental task, which needs governance patterns that are yet to emerge". The main challenge for the stakeholders now is to develop such a sound water regime that establishes the IWRM concept as common rationale for water management in Iran. The products and results developed so far within the German-Iranian project can play a major role in this process. Certainly, the IWRM concept, tools and measures are simply individual pieces of a great puzzle that will only be completed with the decision-makers' strong will to give-and-take.

References

Biswas AK (2004) Integrated water resources management: a reassessment. Water Int 29 (2):248–256. doi:10.1080/02508060408691775

Borchardt D, Bogardi JJ, Ibisch RB (eds) (2016) Integrated water resources management: concept, research and implementation. Springer Publishers, Cham

GWP (2000) Integrated water resources management. TAC Background Papers No. 4, Stockholm

GWP (2003) Toolbox integrated water resources management: sharing knowledge for equitable, efficient and sustainable water resources management, Stockholm

GWP (2009) Lessons from integrated water resources management in practice, Global Water Partnership, Policy Brief 9, Stockholm

Hering JG, Ingold KM (2012) Water resources management: what should be integrated? Science 336(6086):1234–1235. doi:10.1126/science.1218230

Kalbus E, Kalbacher T, Kolditz O, Krüger E, Seegert J, Röstel G, Teutsch G, Borchardt D, Krebs P (2012) Integrated water resources management under different hydrological, climatic and socio-economic conditions. Environ Earth Sci 65(5):1363–1366

Martinez-Santos P, Aldaya MM, Llamas MR (eds) (2014) Integrated water resources management in the 21st century: revisiting the paradigm. CRC Press, Leiden

Molle F, Ghazi I, Murray-Rust H (2009) Buying respite: Esfahan and the Zayandeh Rud River basin, Iran. In: Molle F, Wester P (eds) River basin trajectories: societies, environments and development. Comprehensive assessment of water management in agriculture, vol 8. CABI, Wallingford

Mostert E (2003) The challenge of public participation. Water Policy 5:179–197

The World Bank (2006) Integrated river basin management – from concepts to good practice, Briefing Note 1, Washington DC

UNEP (2012) The UN-water status report on the application of integrated approaches to water resources management, Nairobi

Part I
The Zayandeh Rud Catchment

Chapter 2
Water Resources in Iran and the Reasons for Establishing River Basin Organizations: Review of the First Experience for Zayandeh Rud

Mohammad Ebrahimnia and Bahieh Jafari Bibalan

2.1 Introduction

No one can deny the importance of water in the development of countries or its role in sustaining and preserving human life. Fresh and available water is a natural, rare vital and at the same time renewable resource, which is constantly needed by humans at any time or place. It is also a precious and irreplaceable commodity in the economic and social development of any country, playing a central role in the spatial plans and making up the infrastructure of most sectors. Furthermore it is an important component in the protection, balance and sustainability of ecosystems and the environment, and in particular the water bodies such as wetlands.

Water management in Iran is a complicated affair that includes diverse and numerous structural and non-structural functions. The limitations of available water resources, their inequitable distribution around the country, increasing population and the inappropriate model for urbanization and creation of settlements, as well as the type and method for agricultural production from the point of view of compatibility with climate have not only made water supply a challenge in many parts of the country, but have also intensified its dimensions. Furthermore, the occurrence of consecutive droughts in recent years has made access to water more critical to such an extent that the supply of accessible and safe water for different uses has become a significant challenge and has undermined the socio-economic development of many parts of the country. On the other hand the increased volumes of industrial, urban and agricultural wastewaters and the contamination of water are among the difficulties created in this sector, which increase the sensitivity of the need for coordinated planning.

M. Ebrahimnia (✉) • B. Jafari Bibalan
Ministry of Energy, Tehran, Iran
e-mail: m_ebrahimnia81@yahoo.com

S. Mohajeri, L. Horlemann (eds.), *Reviving the Dying Giant*,
DOI 10.1007/978-3-319-54922-4_2

At first the article introduces briefly the water resources and the climate of the Islamic Republic of Iran and then explains the motives for establishing River Basin Organizations (RBO) in Iran, with a focus on the RBO Zayandeh Rud.[1]

2.2 The Situation of Water Resources in Iran

2.2.1 *Climatic Conditions*

Iran is a large country with special geographical features, and for this reason has a great climatic diversity in different areas. 52% of the surface area of the country is covered by mountains and deserts, and about 16% of the country is located at an altitude of over 2000 m above sea level. While the average annual rainfall at some regions of the central and southern deserts of the Iran does not exceed the 20 mm, the average annual precipitation in the western section of Iran reaches more than 600 mm and in some areas in the northern coastline of Iran it can exceed 1500 mm per year. The average annual temperature at different regions of Iran is also very varied and can fluctuate from 4°C at the Alborz, Zagros and Sabalan heights to 28°C at the coastline of the Persian Gulf and the Oman Sea. Temperatures of above 50°C have also been recorded during summer in some areas in Iran. Moreover at winter, most of the heights have a temperature of −20°C or less. The surface area of Iran exhibits all climatic conditions, from extremely arid to humid, but more than 82% of Iran is located in arid and semi-arid regions.

High precipitation in the Alborz and Zagros mountain chains create humid zones, while the very humid regions are the result of the high rate of rainfall along the coast of the Caspian Sea. The arid and very arid zones are located in the lowlands of the central desert, in parallel to the eastern borderline of the country as well as the coastal regions of the Persian Gulf and the Oman Sea. From the point of view of temperature fluctuations, the country is composed of 17% very cold, 47% cold, 22% temperate and 12% warm areas (Water Research Institute 2014). Figure 2.1 shows the average rainfall fluctuations, while Fig. 2.2 shows the average temperature changes in Iran.

Studies for this paper have shown that in recent years the amount of rainfall has decreased thereby affecting the long-term national average. Figure 2.3 shows the trend of the rainfall fluctuation in different periods.

As shown in the chart, the average rainfall in the recent 15 and 8 years underline the obvious decrease in the long-term and 31-year figures. Moreover, in recent years the average temperature has also increased by 1°C, leading to the reduction of snow cover and glaciers, increased evaporation and consequently the decreased

[1]The organization is actually called Coordinating Council for Integrated Management of the Zayandeh Rud Water Basin.

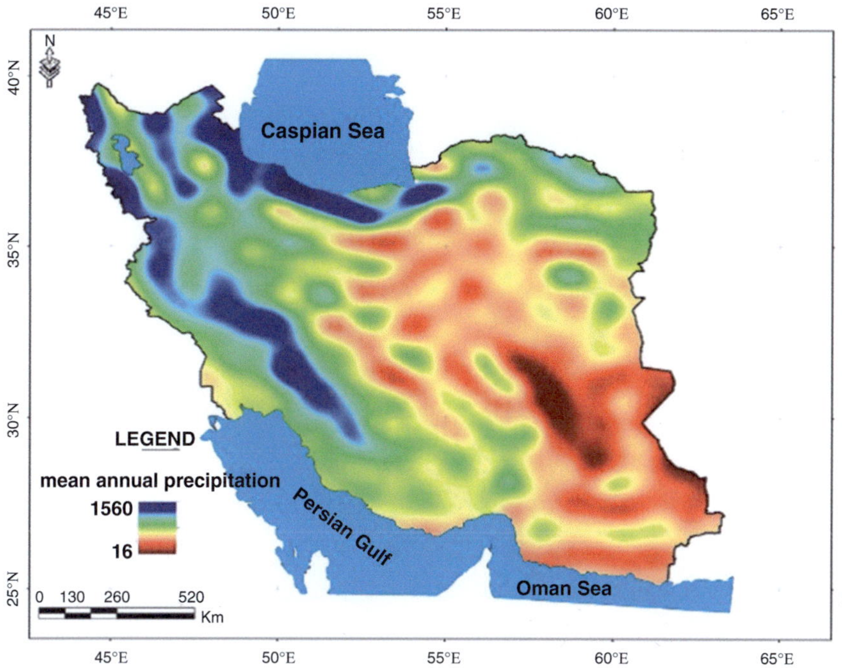

Fig. 2.1 Average rainfall fluctuations in Iran

infiltration and runoff following the rainfall. Figure 2.4 presents the changes in temperature while Fig. 2.5 shows the observed runoff changes in Iran.

In this context the volume of observed runoff has also significantly decreased reaching to the figures of less than 50 BCM in recent years.

2.2.2 The Situation of Renewable Fresh Water

The renewable water resources of any country are the water resources created by precipitation. In fact the volume of precipitation minus evapotranspiration equals the renewable water resource. Based on the volume of rainfall and the surface area of the country as well as the amount of evapotranspiration, the volume of the renewable waters in Iran has also been calculated. Given the reduction of rainfall, the rising temperature and consequently increased evaporation, the volume of the renewable waters in the country has dropped in recent years. Studies show that the renewable water resources of the country amounted to 124.7 BCM before the disastrous break in rainfall, after this break (on average during the past 15 years) it has fallen to 88.7 BCM representing a drop of about 36 BCM. On the other hand, through the increase in population, the per capita renewable water has also dropped by a significant level. Figure 2.6 shows the changes in the renewable waters of Iran.

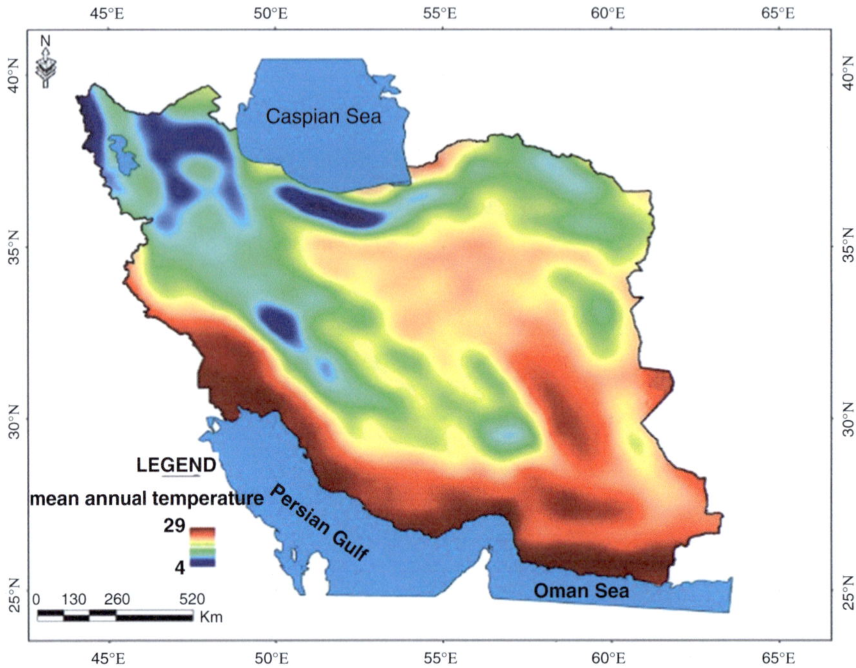

Fig. 2.2 Average temperature fluctuations in Iran

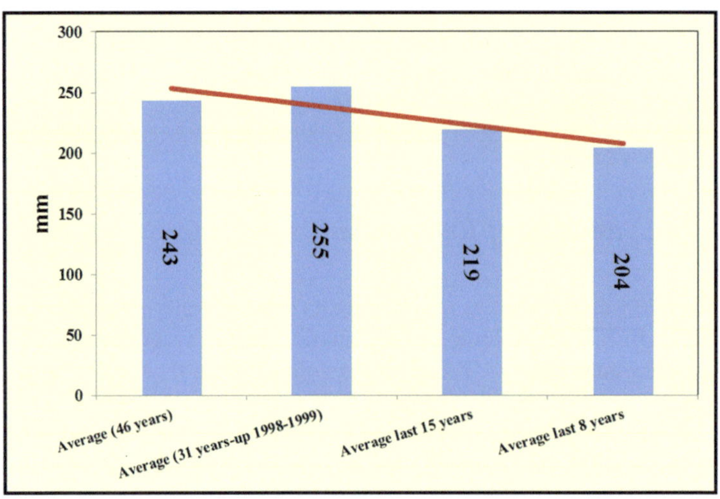

Fig. 2.3 Annual rainfall fluctuation in Iran at different time intervals

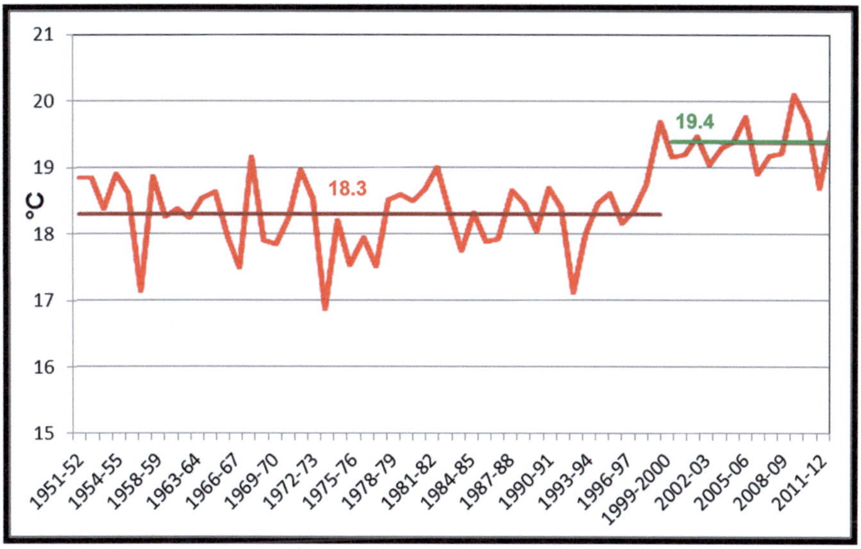

Fig. 2.4 Changes in temperature in Iran at different time segments

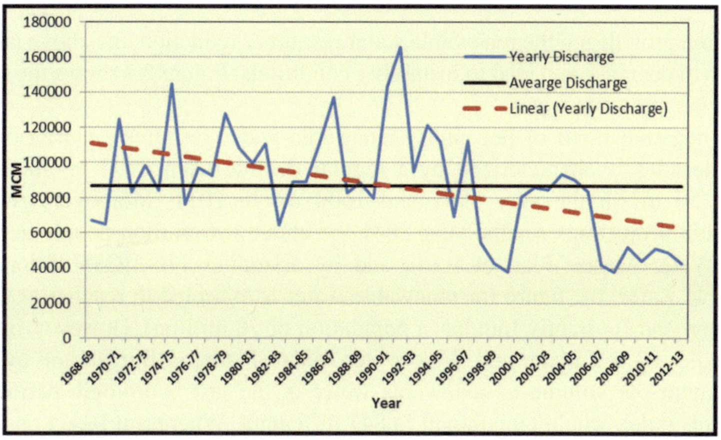

Fig. 2.5 Changes in the long-term observed discharge in Iran

Table 2.1 shows the water stress and water scarcity indicators based on the amount of renewable water resource per capita per year. This table based on the method proposed by Falkenmark and Widstrand (1992) as a result of their study on water resources conditions in many countries. According to Table 2.1, based on the renewable water per capita of the country, the water conditions in an area can be categorized as: no stress, stress, scarcity, and absolute scarcity. The index thresholds 1700 and 1000 m^3 per capita per year are used for water stressed and water scarce areas, respectively (Falkenmark 1989). Therefore, because of continuing

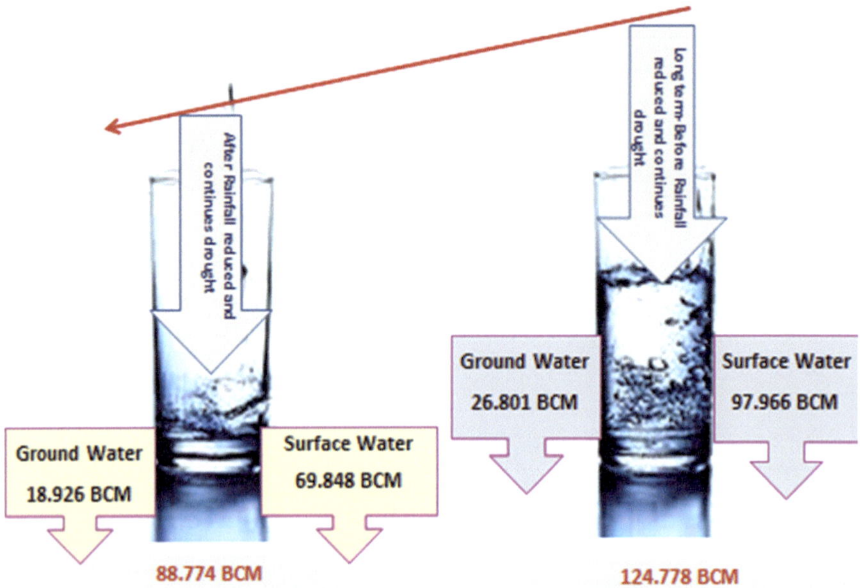

Fig. 2.6 Reduction of natural renewable waters of Iran at different statistical periods

population growth and the renewable water resources reduction, the above indicator in Iran will decrease and lead to instability conditions. Figure 2.7 shows the trend of per capita renewable water.

The negative trend of per capita renewable water resources in Iran over the recent decades is characterized by a regular decline from 6900 m^3 in the year 1956–2156 m^3 in the year 1997 and 1600 m^3 in 2011. Moreover, given the population projections for the year 2041 (of about 106 million people according to the Water Master Plan of Iran), and by assuming 116 BCM of long term renewable water, the figure for renewable water is expected to reach 1099 m^3 per capita per year (currently Iran has a population of 78 million). However, based on new studies by the Water and Wastewater Macro Planning Bureau and by taking into account the volume of renewable water in the last hydrologic period (88.7 BCM) this index would tantamount to 841 m^3/capita. Whereas if based on another studies (Ministry of Energy 2015) assume population of 113 million in the year 2041, this indicator would even decrease to 784 m^3 per capita per year.

Therefore over a period of sixty years the per capita renewable water will change from normal condition (without stress) to the water stress and then towards the limits of water scarcity.

Table 2.1 Water barrier differentiation proposed by Falkenmark (1989)

Condition	Indicator (m^3/capita)
Normal (No stress)	>1700
Stress	1000–1700
Scarcity	500–1000
Absolute scarcity	<500

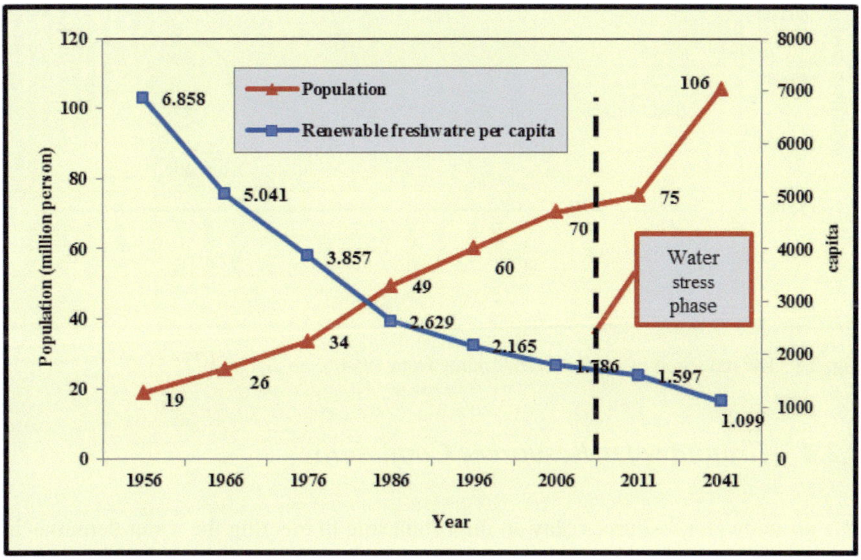

Fig. 2.7 Trend of population increase in and reduction of per capita renewable water in Iran in different years

2.2.3 *Water Consumption*

Despite the renewable water decrease in Iran, the volume of water resources consumption has also increased. The reviewed information underline the fact that according to the data collected for the initial National Water Master Plan in 1999, water consumption was about 88 BCM, whereas during the studies to update the Master Plan (ending on the data of the base year of 2007) it had increased to 100 BCM, of which around 40% are taken from surface waters and around 60% from groundwater resources. It should be mentioned that the shares of agriculture, domestic and industries from these 100 BCM are, in order, 92%, 6.5% and 1.5% (Ministry of Energy: Updating of Water master plan of Iran 2007). Whereas, the consumption rates according to the world average are 70% in agriculture, 20% in industries and 10% in households. It can be observed that compared to the world average especially agricultural consumption in Iran is extremely high and requires special attention in water consumption management.

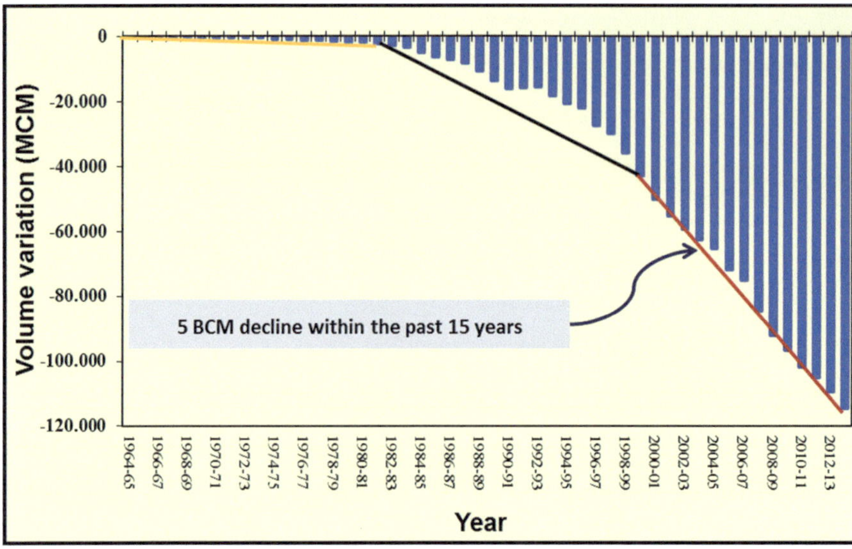

Fig. 2.8 The reduction of groundwater volume from 1964/65 to 2013/14

2.2.4 Groundwater Resources Conditions

The groundwater resources play an important role in meeting the water demands in Iran, to the extent that 57% of urban potable waters, 83% of the rural potable waters, 63% of industrial and service waters and 52% of agricultural waters are supplied from these resources (Ministry of Energy: Updating of Water master plan of Iran 2007). The ease of extraction from these resources and their cheap price has led to their uncontrolled exploitation (See Fig. 2.8).

Currently, from the 609 studied zones (plains) in the country, only 289 have unlimited withdrawal rates while the remaining 320 zones are restricted for water exploitation. Anyway, a considerable number of large and plains with important groundwater resources have been destroyed to the extent that 78% of these are have been declared restricted zones. The overexploitation of these resources has resulted in their dramatic drop. Today, the aggregated surplus extraction of groundwater amounts to 110 BCM. The aggregated drop in the groundwater tables during the statistical period is about 15 m (see also Chap. 15 Sklorz). Continuous depletion of aquifers has led to their mortality, and while the water yield of wells does not increase any more or even drops, the number of wells increases as result. This fact is clearly shown in Fig. 2.9. As pointed out earlier, the volume of water consumption in Iran is heavily dependent on groundwater. Over the last 40 years the volume of consumption from groundwater has quadrupled, while the number of wells over the same period has increased 16 times.

The dependence of water supply on groundwater resources in Iran means that the resulting depletion of these resources and their quality deterioration has created

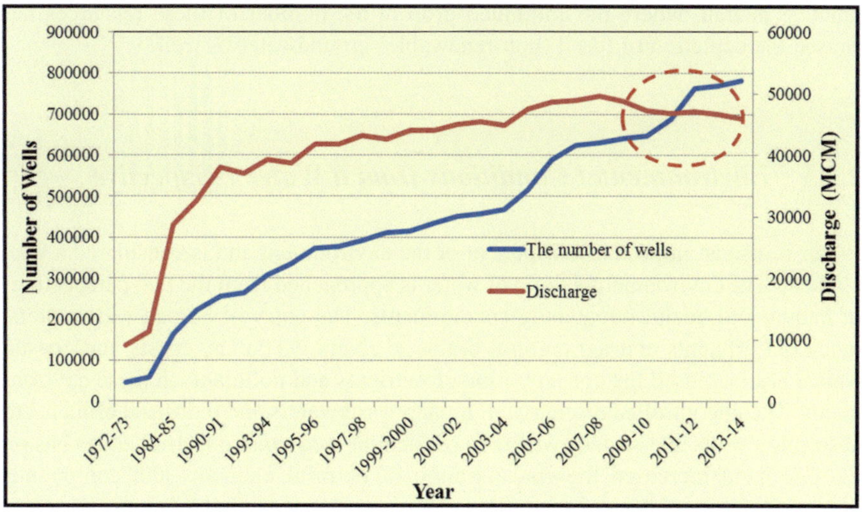

Fig. 2.9 The trend of number of wells and the reduction in their discharge

many difficulties for potable water supply, especially in rural areas. The potable water of most villages is supplied from springs, qanats[2] and/or wells, but due to current reduction in rainfall, extraction from groundwater resources exceeding their renewability, and soil impermeability, these resources have either dried up or have witnessed a deterioration of their quality.

Agriculture's dependence on groundwater leads to a serious situation. The continued drop of groundwater tables in some plains coupled with a lack of water supply has resulted in a lot of arable land that remained unused and many years-old horticultural gardens are on the path of destruction. Examples can be seen in the provinces of Kerman and Fars as well as in many other parts of Iran. As far as industry goes, the shortage of water resources has undermined the trend of industrial development in many regions. It is not possible to allocate groundwater in most areas and consequently the industries are obliged to buy agricultural wells thereby creating further problems and difficulties. Land subsidence is another consequence of the drop of water levels in aquifers and their over-exploitation. In addition to creating hazards for infrastructures, by changing the land morphology, altering the form of particle settlement and by reducing soil porosity, land subsidence thwarts attempts to replenish and revive aquifers.

In many countries of the world groundwater resources are counted among the strategic water resources. Efforts are made to preserve these resources for future generations and to use them only during crises. This however, is not the mode of

[2]Artificial underground channel with vertical access shafts, used to transport water from an aquifer under a hill.

practice in Iran, where the continued drop in the number of these resources has caused the depletion of fossil (non-renewable) groundwater as well.

2.2.5 Environmental Conditions from a Water Perspective

Water is also an important component of the environment and is considered a vital element. The environmental topic of water is approached from the two perspectives of habitat and health threatening contaminants. The entry of human wastewaters, industrial effluents and agricultural drains to rivers, wetlands, coasts and coastal waters has increased the concentration of nutrients and pollutants in these environments. The uncontrolled discharge of human wastewaters and industrial effluents to absorption wells and surface waters in residential areas and industrial zones has on the one hand increased the concentration of harmful elements and compounds beyond the permitted environmental standards, and on the other hand, the improper use of such effluents in agriculture may ultimately create public health problems after the consumption of affected crops. The increased salinity of groundwater resources is another problem in the sector.

Water resources play an important role in the national water ecosystem of Iran. There are 95 important wetlands in Iran, 35 of which are registered under 24 titles in the Ramsar Convention, while the 65 wetlands unregistered in the above convention are managed by the Department of Environment under the title of four regions. The occurrence of consecutive droughts and the unsustainable use of water resources have dried up many wetlands. The Urmia Lake, the wetlands of Tashk, Bakhtegan, Hamoon and Parishan, the Zayandeh Rud River and many others are examples of environmental problems created by water scarcity, which will definitely leave a negative impact of public health as well.

It can therefore be concluded that water has an essential and pivotal role in the creation of a sustainable ecosystem and in establishing the principles of sustainable development, and if there are no regular and integrated plans for water management, the water crisis in Iran will undermine essential infrastructures.

In this context, given the consecutive droughts as well as the contamination beyond the capacity of basins, the instability of many catchment basins has obviously led to problems for the livelihood of many people. In the face of these challenges, the Ministry of Energy established the Zayandeh Rud River Basin Organization in 2014.

2.3 Water Resources Management in Iran

2.3.1 Water Management Challenges in Iran

Based on the described water crises in Iran, water resources management and the water sector as a whole face challenges as summarized below:

- Climate change, droughts and reduction of renewable water resources;
- Lack of a coordinated spatial plan for the development of industries, urban and rural settlements, and agriculture;
- Lack of economical view on water and the low attraction of water sector for private investment;
- The low productivity of water in the different consumption sectors;
- Imbalance of water supply and consumption and increasing water competition (between users) and the disproportionateness of the master planning in the execution of projects to develop water resources and agriculture based on national water capacities in the river basins;
- Deterioration of water quality and pollution of water resources;
- Lack of integrated management of water resources at river basin level that would constitute the most appropriate scale for water resources management;
- Limitations of comprehensive planning for water within the geopolitical boundaries of the provinces and the availability of political and social pressures to obtain a greater share of water resources and the belief in the culture of haggling during the study and execution of provincial plans; and
- Instable environment of the river basins as an ecosystem unit.

2.3.2 The Structure of Water Resources Management in Iran

2.3.2.1 The Trend of Water Structure Formation in Iran

Following the global trend, the existing system of water resources management in Iran was formed gradually during the last 80 years. During this period, the population of Iran has increased seven folds and the ratio of urban to rural population is now weighing heavily on the urban side. Therefore, the trend of formation of the national water resource management structure can be divided in the following distinct periods (see Nikravesh 2010):

The Initial Stage This period starts at the beginning of the twentieth century, i.e. around the year 1927, when the population in Iran was around 10 million, and continues until the year 1963. At the beginning of this period and prior to it, the water sector had no place in national organization and planning, and the land owners addressed and resolved the local water management issues within the feudal system frame. In this period, the special Law on Qanats was approved in the year

1930, the law on the establishment of an independent irrigation agency, which was mainly responsible for creating irrigation and drainage networks, was enacted in the year 1943, and the law for executing the water and wastewater pipeline in the city of Tehran was ratified in the year 1951.

The Formation Stage At the beginning of the 1960s, Iran was on the threshold of social and economic evolution, which led to the abolishment of the feudal system. This stage included the expansion of urbanism and services with the higher concentration of power in the hand of the central government. In this period, which extended from the year 1963–1978, the law for the establishment of the Ministry of Water and Energy was approved in the year 1963; the law for conservation and protection of the national groundwater resources in 1966, the law for water and the procedure for its nationalization in 1968 and the law for the establishment of the Ministry of Energy was ratified in 1975.

The Stage of Growth This stage began in the year 1978, i.e. after the victory of the Islamic Revolution, and was characterized by a penchant for the creation of water structures, in particular of dams. Although at the beginning there were ambiguities about the duties of water management in Iran, later on a division was set between the water and power bodies. Among the most important actions taken in this period, one may cite the ratification of the law for equitable distribution of water in the year 1982, the first law for the economic, social and cultural development of the country in the year 1988, the law to stabilize the agricultural water price and the law for the establishment of water and wastewater companies.

The Stage for Improvement and Integration This ongoing stage has begun approximately at the end of the 1990s when the issue of integrated management of water resources at the river basin level was raised. During this period many intellectual actions were taken, but still the penchant for creation of large water infrastructures remained the main approach. Among the most important actions taken in this period are; preparation and communication of general military policies on water, long-term development strategies for water resources, preparation of legal act of water section vision. In fact the focus in this period was to present an integrated and holistic policy for water resource management in tune with the global society and new ideas.

Accordingly, at present the most important missions and the main governance tasks related to water are among the mandates of the Ministry of Energy (the departments of Water, and Water and Wastewater Affairs) and according to the division, the administrative tasks are carried out by the two specialized holding companies, the Iranian Water Resources Management Company (IWRMCo.) and the National Water and Wastewater Engineering Company (NWWEC) as well as their subsidiaries. Of course, some of the governance tasks are also undertaken as brokerage by the tow holding companies. In fact issues related to water resource management are handled by the Water Resources Management Company and the task of supplying potable water and wastewater services are carried out by the National Water and Wastewater Engineering Company. Each of these holding

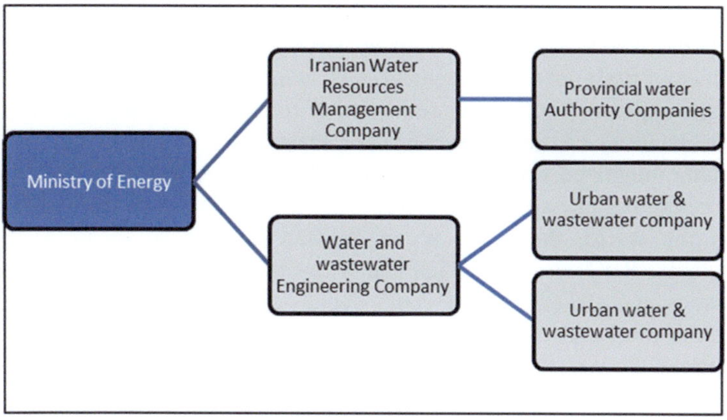

Fig. 2.10 Structure of the water and the water and wastewater sectors of Iran

companies have one subsidiary in each province, and unfortunately, despite the fact that the water resource management ought to be undertaken at river basin level, in Iran the political divisions (provinces) are in charge of these affairs. Of course, in the context of implementing water resource management at basin level, the bureaus for management of catchment basins have in recent years been established in the Water Resources Management Company of Iran, which include the Bureau for Integrated Management of the River Basins of the Central Plateau, Ghareghoum and the eastern border; the Bureau for Integrated Management of the River Basins of the Caspian Sea and the Urmia Lake and the Bureau for Integrated Management of the River Basins of the Persian Gulf and the Oman Sea. The structure of the national water sector is shown in Fig. 2.10.

2.3.3 The Reasons for Establishing River Basin Organizations in Iran

Given the multi-disciplinary nature of water and its impact on different consumption sectors, and based on the law for the establishment of the Ministry of Agriculture, the Supreme Water Council was founded. It is headed by the President or the Vice President and composed of the Ministers of Energy, Agriculture, Industry, Mining and Commerce as well as the heads of the Department of Environment and the Management and Planning Organization, an observer from the parliament and two water experts, to coordinate the policymaking in the fields of water supply, distribution and consumption. This council constitutes the highest level of decision making about water in Iran.

It should be mentioned that following the establishment of provincial companies to manage water resources, tensions about the extraction of the common water resources have increased, and each province, regardless of the capacity of the basin's water resources, started to set up water resources development plans. In the absence of basin level management, this situation has led to water withdrawal beyond the capacity of the water resources of river basin, with the occurrence of droughts, intensifying these tensions. For these reasons and in the context of establishing integrated water resources management, the lack of a basin management structure became apparent. Accordingly, after a review of the existing structures in the world, the water sector in Iran decided to create river basin organizations (RBO) through which and with the collaboration of all the basins' beneficiaries and stakeholders it could decide about the management of water resources in the basin. The idea for establishing and proposing RBOs arose for the first time following the intensification of problems and conflicts in the Zayandeh Rud river basin, and it was approved during the 10th session of the Supreme Water Council. Based on the experiences of this council, the RBO establishment for other basins was also approved during the 16th session of the Supreme Water Council. RBOs should be composed of the Minister of Energy (or his fully authorized representative) acting as the RBO's president and the governors of beneficiary provinces located in the basin, the deputies of the Ministers of Energy, Agriculture, Industry, Mining and Commerce, Interior Affairs, the Department of Environment and the Management and Planning Organization and the representatives of farmers in the relevant provinces.

2.4 The Importance of the River Basin Organization for the Zayandeh Rud River Basin

2.4.1 The Specifications of the Zayandeh Rud River Basin

The Zayandeh Rud River originates in the eastern flanks of the Zagros heights and the high mountains of Zardkooh Bakhtiari, and after passing the 300 km of river course ends up in the Gavkhuni wetland. This basin extends over an area of 26,917 km^2 and is divided between the provinces of Esfahan and Chaharmahal and Bakhtiari as shown in Table 2.2. The Zayandeh Rud River is the principal river of the Gavkhuni Basin.

Table 2.2 The division of Zayandeh Rud Catchment between the two provinces of Esfahan and Chaharmahal and Bakhtiari

	Chaharmahal & Bakhtiari	Isfahan	Total
Basin's surface area (km^2)	1912 (7% of total)	25,005 (93% of total)	26,917
Population in 2011 (persons)	70,771 (1.8% of total)	3,922,609 (98.2% of total)	3,993,380

The long term average rainfall of the Gavkhuni basin is equal to 268 mm. Over the last 15 years, the Zayandeh Rud River has experienced two drought periods, one lasting for four and the other for 7 years. In the first period, which lasted from 1998 to 2002, the Zayandeh Rud dried up for the first time within the boundaries of Esfahan after the construction of the Zayandeh Rud reservoir dam. The second time, which began in the year 2007 and continues on today, the average annual discharge of the Zayandeh Rud River has decreased from 1500 MCM/year to about 1000 MCM/year. Since in addition to the supply of water to about 200,000 hectares of agricultural land, the river has to meet the potable water demands of a population of 5 million people in three provinces of the country (Esfahan, Yazd,[3] and Chaharmahal and Bakhtiari) as well as the demands of regional industries. Therefore, in the light of drought conditions, the water resources management of Zayandeh Rud has faced many difficulties and challenges in recent years. According to the existing data and documentation, from the beginning of the Zayandeh Rud Dam's operation until 1998 (before the first drought period), due to the large volume of river discharge, an adequate volume of water even in excess of their water rights was distributed among the farmers of Esfahan Province. However, like in any other part of the country, since 1998 to date, subsequent to two consecutive droughts, today it has become difficult if not impossible to fully meet the farmers' claims according to their water rights.

In the recent 59 years, where reliable data is available for the Zayandeh Rud river basin, there were 26 wet years against 25 years of droughts, which are still continuing. The review of hydrological periods and the historical witnesses and documents confirm that the phenomenon of drought in the Zayandeh Rud Basin has been a continuous feature from centuries ago to date, leading to the conclusion that drought is a natural phenomenon of Zayandeh Rud climate. Tables 2.3 and 2.4 show the drought conditions in recent years and the historical references on Zayandeh Rud Basin.

For an insight in the condition of water entering the Zayandeh Rud Dam, the amount of water entering the dam from the water year of 1991/92 until the year

Table 2.3 Drought condition in the recent century

Condition	Length of the period (in years)	Date
Drought	11	1956–1967
Wet	15	1967–1982
Drought	3	1982–1985
Wet	11	1985–1996
Normal	3	1996–1999
Drought	3	1999–2002
Normal	5	2002–2007
Drought	8	2007–2015

[3]Water is transferred to Yazd Province through artificial canals.

Table 2.4 Important historical droughts and their consequences

Date (Hijjara)	Name of historian	Consequence
423	Nasser Khosro	Great famine–Cholera epidemic
1130	German historian	Famine forcing people to eat carcass
1165	Jean Guoret	Loss of horticultural gardens
1288	Abdollah Mostowfi	Death of thousands of people
1335	Dr. Hassan Hosseini Abri	Intense famine–social uprising
1366	Mohammad Hassan Khan Jaberi	Death of thousands of people

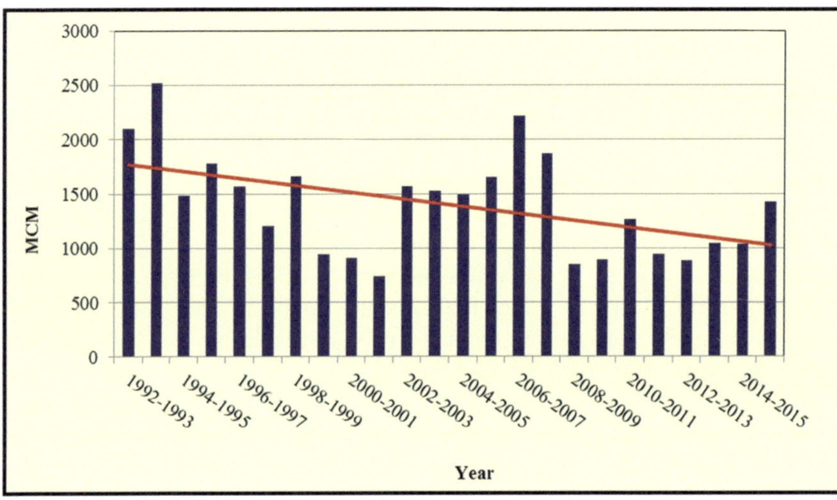

Fig. 2.11 Volume of water entering the Zayandeh Rud Dam between 1992/93 and 2014/2015 (MCM)

2015/16 is shown in Fig. 2.11. Studies show that the volume of water entering the dam has declined.

2.4.2 *The Challenges in the Zayandeh Rud River Basin*

The uncontrolled expansion of industrial and urban centers (including population growth) in the region in recent years, the increasing water demand of these centers and the reduction in the annual flow of the Zayandeh Rud River have undermined the balance between water resources availability and consumption in the Zayandeh Rud river basin. Moreover, undisciplined water withdrawal and pollution in the basin, the need to supply potable and industrial waters and to meet the water demands of the agricultural and environmental sectors in the Zayandeh Rud river basin have put the Ministry of Energy and other relevant regional and national authorities in a very difficult situation of having to solve these issues. The general features of the Zayandeh Rud river basin can be described as follows:

- The most sensitive catchment in the country from a political and social point of view;
- Supplier of potable water to a population of over 5 million in three provinces of Esfahan, Yazd and Chaharmahal and Bakhtiari;
- Supplier of agricultural water for over 200,000 hectares;
- Supplier of water to large industries such as the steel mill, Mobarakeh Steel, power plant, refinery, Sepahan Cement and poly acryl etc.
- The hub of tourism in the central plateau.

On the other hand the challenges of the mentioned basin are:

- Imbalance between the water resources of Zayandeh Rud river basin and the extensive consumptions in the agricultural, household and industrial sectors and the uncontrolled and unsustainable development in the region;
- Unconventional competition among the provinces of the basin, the incitation of public opinions and the intensification of the social gaps;
- Excessive pressure on the river basin beyond its capacity along the 400 km course of the river from its source to estuary;
- Continuation of drought and the critical diminution of the Zayandeh Rud Dam's reserves;
- Lack of solidarity and cooperation among the authorities and beneficiaries of the catchment basin in addressing the problems; and
- The high susceptibility of the region to social unrest and tensions.

Therefore, given the sensitive situation of this basin, the differences created between the two provinces of Esfahan and Chaharmahal and Bakhtiari, the competition for greater share of the basin's water resources and the deteriorating condition of the basin's environment, the Ministry of Energy decided to address the issues and problems of this basin at the highest management level and to create the necessary accord and coordination in this regard among the beneficiaries and the stakeholders. In this context the issue was discussed during two meetings of the Supreme Water Council, and according to its ratifications, in addition to the prohibition on new water extraction permits in the basin and clarifying the status of illegal withdrawals, it was decided to establish the river basin organization for the Zayandeh Rud river basin and thus, implementing integrated water resources management (IWRM).

With the objective of implementing IWRM in the Zayandeh Rud river basin and reducing the conflicts on the water resources of this basin, the RBO was established in the Ministry of Energy, with the Minister of Energy acting as its President and the governors of Chaharmahal and Bakhtiari, Esfahan and Yazd, the deputies of the Ministers of Energy, Agriculture and Industry, Mining and Commerce and two farmers' representatives of the provinces of Chaharmahal and Bakhtiari and Esfahan making up its members.

2.4.3 *The Results of Forming the RBO Zayandeh Rud River Basin*

Given the intense conflict among the two beneficiary provinces over the basin's water resources, the volume of extractions and the execution of two water transmission projects (Golab 2 and Beheshtabad), the initial meeting of the RBO was devoted to the issues of finding the accurate number of resources in the basin and the volume of water extracted for different uses in the two provinces; review of the water transmission plans and their necessity and creating conditions for their implementation. Therefore, in the second meeting of the RBO, the picture of resources and water consumptions in this basin was finalized and it was the beginning of the IWRM project implementation. After raising the subject of this board during the 13th session of Supreme Water Council and given the protests of the two provinces against the decisions proposed, especially about the water transmission plans, it was decided to review the issues once again during different meetings together with the political authorities of the provinces. Finally, they were presented and approved during the fourth session of the RBO. The approval of the board, however, meant the stoppage of some development plans in the two provinces, and provoked strong political opposition in both provinces. To date the RBO has held nine meetings, which created a certain level of consolidation in the provinces, but the RBO's decisions have not been fully implemented yet. For instance, regarding the intermission of the development plans, despite requests to the concerned provinces through different notifications and letters exchanged at various levels, the plans have not yet come to a complete halt. In spite of the emphasis on the necessity for the implementation of the transmission plans to meet the potable water demands in the provinces of Esfahan, Kerman and Yazd, and for changes in the method of conveyance and separation of the lines, the members of parliament in the province of origin still criticize the mentioned plans under the pretext of illegality or non-professionalism to create assurance in the beneficiary provinces. Although as the first step and the first experience in field, the RBO has partially succeeded in improving the water resource management at basin level, it is still far from ideal.

Some of the current challenges of the RBO are:

- Focus of provinces on water resources and consumption and demanding greater share of the basin instead of concentration on solutions for optimum use of the allocated water. To date the expected quantity and quality objectives of IWRM in the basin have not been met.
- Focus on information related to the water sharing and allocation and the lack of detailed and accurate IWRM studies, especially in the field of water consumption (in some cases such as the allocation of 200 MCM of treated effluents to industries as per the decision of the RBO, without undertaking detailed and comprehensive studies on the subject).

- Incomplete implementation of the RBO's decisions, especially those related to stoppage of new and water demanding development plans.
- The existence of a competitive atmosphere among the two beneficiary provinces, the preference for provincial/sector benefits over national interests and self-righteous attitude in the greater and unsustainable use of the basin's water resources instead of their sustained and optimized exploration.
- Lack of coordination among the decisions of relevant bodies in the basin on execution of development plans (despite the RBO's decisions, some organizations still pursue the aim of implementing new development plans).

It should be mentioned that along with the Bureau for Integrated Management of the basins of Central Plateau, Ghareghoum and Eastern Border basins, the RBO Zayandeh Rud is affiliated to the Iranian Water Resource Management Company. The manager of the basin along with provincial water authorities are in fact the executive of the RBO's decisions and monitor their implementation. By hiring two qualified consultants, establishing technical and executive agents in the Zayandeh Rud Basin (in the towns of Saman and Chadegan) and constant field presence, the basin's management has realized the permanent monitoring of water extractions and the managing of the water resources of the Zayandeh Rud basin. Moreover, it has established direct contact with the beneficiary groups and has organized different and periodical meetings with the stakeholders, beneficiaries and other specialists. Furthermore, the two agents are responsible for impartial and fair monitoring and handling of major problems of the Zayandeh Rud basin among the beneficiary provinces. To better pursue the issues of the basin, establish a close relation with the beneficiaries and to keep a track of the intakes, a working group on the basin's water resources and consumption was also formed with the participation of representatives from relevant organizations and beneficiaries. Its objectives are to collect accurate details of water resources and consumption, and to assure between the two provinces and relevant stakeholders for implement the RBO's decisions.

2.5 Conclusion

By studying similar structures for RBOs in different parts of the world, it appears that the RBO for the Zayandeh Rud basin has still not found its real status in the management of the basin and having an impact on it. In most RBOs, decisions-are taken at the lowest appropriate level. Whereas under the current conditions in the Zayandeh Rud RBO the final decision maker is the central governmental, although through the presence of farmers' representatives (one representative for each province), efforts were made to incorporate the views of the main beneficiaries in these decisions. Of course in other parts of the world the evolution process is completed and the current culture has been achieved through time and in gradual steps. Iran still lacks the necessary maturity to take decisions without the

Table 2.5 List of rivers prioritized for establishing RBOs (Ministry of Energy 2016)

Grade 1 Catchment Basin	Name of River/ catchment basin	Relevant provinces
Caspian Sea	Aras	West Azerbaijan/East Azerbaijan/Ardebil
	Atrak	Razavi Khorasan/North Khorasan/Golestan
	Gorganrood	North Khorasan/Golestan/Semnan
	Haraz	Mazandaran/Tehran
	Sefidrood	Kurdistan/Hamedan/Zanjan/East Azerbaijan/ Ardebil/Guilan/Alborz/Qazvin and (Tehran)
Urmia Lake	Urmia Lake	West Azerbaijan/East Azerbaijan/Kurdistan
Persian Gulf and Oman Sea	Great Karoun	Esfahan/Chahar Mahal and Bakhtiari/ Kohgiluyeh and Boyer Ahmad/Markazi/ Lorestan/Khuzestan and (Qom)
	Karkheh	Hamedan/Kermanshah/Kurdistan/Ilam/ Lorestan/Khuzestan
	Sirvan Sardsiri	Kurdistan and (Kermanshah/Ilam)
	Zohreh	Fars/Kohgiluyeh and Boyer Ahmed/Khuzestan and (Bushehr)
	Jarrahi	Kohgiluyeh and Boyer Ahmed/Khuzestan
	Zab	West Azerbaijan/Kurdistan
Central Plateau, Ghareghoum and East border	Zayandeh Rud	Esfahan/Chahar Mahal and Bakhtiari and (Yazd)
	Karaj	A;borz/Tehran
	Hablerood	Tehran/Semnan
	Ghomrood	Markazi/Qom
	Maharloo—Bakhtegan	Fars
	Kashafrood	Razavi Khorasan

dominance of certain government officials and this exercise needs to be practiced. Still the individuals present in the RBO's meetings supersede their own interests over national or basin-wide ones, and a change of attitude in this respect requires exercise and repetition.

Moreover, the plans to form RBOs for the river basins of 18 prioritized rivers are under way. These rivers, including the Zayandeh Rud, are listed in the following Table 2.5:

As pointed out in this chapter, given the current conditions including reduced rainfalls, increased temperature, decreased runoffs, illegal explorations in the basin from surface and ground waters, the problems of quality and the environmental issues, the social and economic problems created by different means, by establishing RBOs and by engaging participation of beneficiaries and stakeholders alike, the water sector of Iran tries to address IWRM as a priority in the problematic catchment basins; and in this way, in addition to ensuring the sustainability of the territory, it can create the necessary balance between available water resources and consumption. Of course the social and cultural issues are among problems that deserve serious attention in this context, because without a social and cultural support, the creation of joint decisions and mandating the implementation of the RBOs decisions, one cannot expect the accomplishment of its goals.

References

Water Research Institute (2014) Iran's water feature, present condition. Tehran

Falkenmark M, Widstrand C (1992) Population and water resources: a delicate balance. Population bulletin, Population reference bureau

Falkenmark M (1989) The massive water scarcity threatening Africa - why isn't it being addressed. Ambio 18(2):112–118

Ministry of Energy: Updating of Water master plan of Iran, 2007–2014, Water and wastewater macro planning bureau, Tehran

Ministry of Energy (2015) Studies of water allocation commission, Water and Wastewater Macro Planning. Bureau, Tehran

Ministry of Energy (2016) Iranian water resources management company, Tehran

Nikravesh M (2010) Institutional Studies on Water Resources Management in Iran concluding to introduction of an efficient structural pattern, Applied Research. Iran Water Resources Management Company, Tehran

Chapter 3
Water Management in the Zayandeh Rud Basin: Past, Present and Future

Mohammad Ali Torfe, Masoud Mirmohammad Sadeghi, and Shahrooz Mohajeri

3.1 Characteristics of the Zayandeh Rud Catchment

The Islamic Republic of Iran is divided into six main hydrological basins: the Central Plateau, the Persian Gulf and Oman Sea, the Caspian Sea, the Urmia, the Ghareghoum, and the Eastern Boundary basin. These basins, again, are divided into 30 main catchments with different sub-catchments. The Zayandeh Rud sub-catchment belongs to the Gavkhuni main catchment which is located in the Central Plateau of the Iran basin (Fig. 3.1).

The Zayandeh Rud, the "life giving river", is the most important surface water in the Central Plateau of Iran, a typical (semi-) arid desert. It originates in the Zagros Mountains in the Chahar-Mahal and Bakhtiari province through a natural merger of small and large rivers at an altitude of around 4200 m. The largest part of its 26,000 km^2 catchment spreads out downstream in the Isfahan Province. On its 405 km course, the Zayandeh Rud runs through extremely different climatic and natural conditions with, as a consequence, various socio-economic characteristics. (Shafaghi 2003; Hossaini Abari 2000, more details in Chap. 14 Faramarzi) The existence of deep and very fertile soils, particularly silts and clay loams, has led to the development of intense agricultural activities along the river (Sarhadi and Soltani 2013; Molle et al. 2009).

The catchment can be divided into three main parts: the (sub-) alpine region and the foothills, the Isfahan lowlands including the marshland and contiguous desert

M.A. Torfe
Governor's Office of Isfahan Province, Isfahan, Iran

M. Mirmohammad Sadeghi (✉)
Isfahan Regional Water Company, Isfahan, Iran
e-mail: msadeghi84@yahoo.com

S. Mohajeri
inter 3 GmbH Institute for Resource Management, Berlin, Germany

S. Mohajeri, L. Horlemann (eds.), *Reviving the Dying Giant*,
DOI 10.1007/978-3-319-54922-4_3

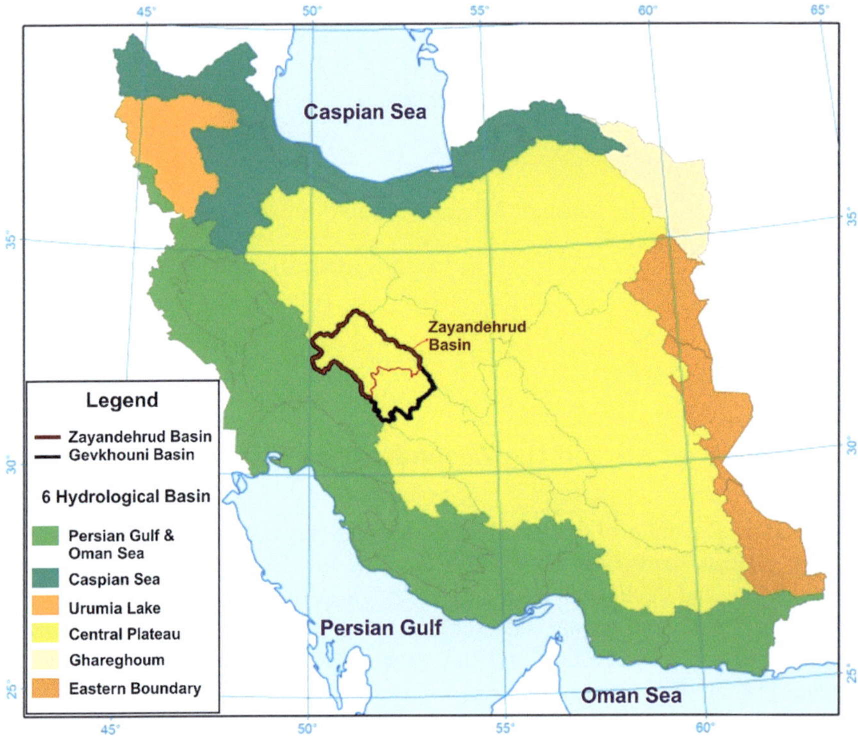

Fig. 3.1 Position of Gavkhuni and Zayandeh Rud basins in relation to the 6 main hydrological basin of IRAN (Isfahan Water Authority

(Fig. 3.2). The first region in the western part of the catchment shows climatic conditions typical for Central Asia's high-altitude mountains, with very cold days as low as −30 °C and heavy precipitation. Average annual precipitation is around 1260 mm, mostly in the form of snow, leading to a permanent snow cover of up to 7 m in higher regions (Fig. 3.3). Only in summer (July to September) is there no precipitation. Below the timber line (approx. 2300 m) a mountain forest steppe with oaks and a diverse fauna – like brown and black bears and other large mammals – can be found. Grazing pressure, timber extraction and pruning have led to a retreat of forests and an expansion of thorn hedges and brushwood (Sahafii and Sadeghi 1997). In this area, around 100,000 people live on agriculture (stock and arable farming) and forestry under difficult conditions, and for years there has been a heavy rural exodus, to Isfahan city in particular.

The foothills are located in the Isfahan province. Here, the Zayandeh Rud dam is located, with its reservoir of around 1400 MCM as the main source of water for the entire catchment (Sarhadi and Soltani 2013). The foothills' wide landscapes form the transition to semi-arid regions with around 430 mm of precipitation and snowfall in winter. In addition to oak forests, almond trees and agriculturally cultivated areas can be found. Around 250,000 people live in villages and smaller

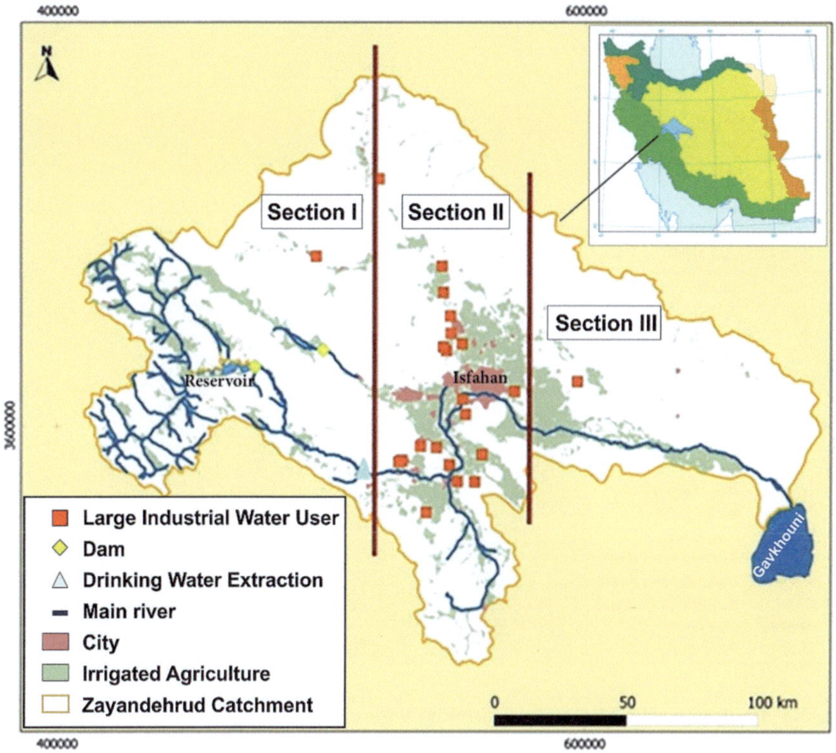

Fig. 3.2 Three main parts of Zayandeh Rud catchment and the location of main water users

towns and work mainly in agriculture but also in small craft and industrial enterprises. In the agricultural areas primarily wheat, almonds, potatoes and rice are grown. Molle et al. (2009) point out that today irrigated nut and almond orchards located in the Chahar-Mahal and Bakhtiari province are expanding, while traditionally, irrigation was restricted.

In the first part the water flows in what can be described as quite good quality around 100 km from the Zayandeh Rud dam to the Chamaseman dam, the extraction point of Iran's biggest water works (Mohajeri and Dierich 2008). During the last few years, the region around the Chamaseman reservoir has become a famous recreation destination. Here, the drinking water for the 4.5 million citizens of the Isfahan province and other cities outside of the catchment area (such as Yazd, Kashan or Naein) is extracted.

In the elongated Isfahan lowlands, annual precipitation decreases from around 200 mm in the western part to 85 mm in Isfahan city and its surroundings (Fig. 3.3). This part of the catchment, a flood plain with fertile soils ranging around 150 km to the east and west of Isfahan city, is used for intensive agriculture, mostly of staple foods and fodder (Molle et al. 2009). As a result of intense cultivation and irrigation measures, agriculture shapes the actual grass land and steppe landscape. In this part

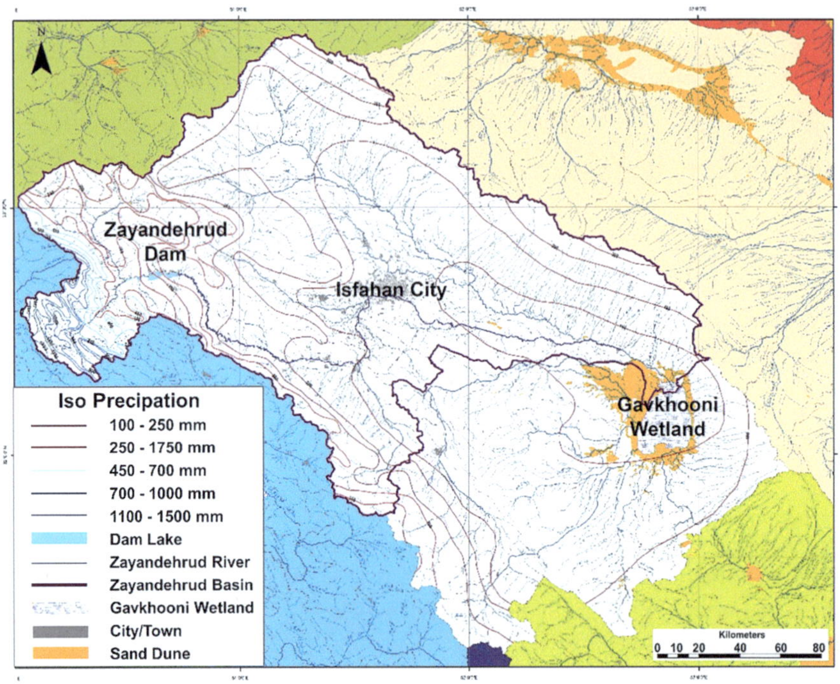

Fig. 3.3 Iso precipitation Map of Zayandeh Rud Basin

of the catchment, most small and big towns and villages were founded in a long track along the river, and most inhabitants of Central Iran live in this area (Hossaini Abari 1991). Isfahan is Iran's third biggest city, and its historical buildings were declared UNESCO World Heritage. During the last decades, the region became Iran's largest industrial area, with mainly ironworks, steel industry, oil refineries, and chemical industry. Here the river water is heavily polluted through waste water and agricultural drainage water.

Marshland and desert east of Isfahan city form the last large area in the catchment. Annual precipitation is only around 55 mm, and with heavy winds and frequent sand storms the region belongs to the arid parts of the country. The character of the ecosystem depends on the distance from the Zayandeh Rud River: In the (former) flood area of the river is marshland which turns into Artemisia Steppe and desert further away from the river. Due to irrigation measures, onions and other vegetables can be cultivated close to the city of Isfahan, while further downstream primarily cotton is grown. Climate change has been forcing more and more of the 200,000 inhabitants to leave this area.[1] Around 125 km southeast of Isfahan, the river flows into the Gavkhuni salt lake, a marshland internationally

[1]Particularly in summer, water shortages lead to problems regarding drinking water supply. Many people therefore move to cities during this time.

acknowledged by the Ramsar Convention in 1975 and important for staging and wintering for several species of migratory water birds.[2] Moreover, it forms an important barrier against the desert. Figures on how much water the Gavkhuni wetland requires to be a functionally healthy wetland ecosystem range from 70 MCM (Salemi and Murray-Rust 2002) to 243.5 MCM (Sarhadi and Soltani 2013). Over the last 15 years, however, neither of these amounts has been met, and the water arriving at the Gavkhuni Lake has been almost exclusively agricultural drainage water.

3.2 Increasing Water Demand

20 years ago there were "serious concerns that it will be extremely difficult, if not impossible, to meet expected demands for water for Isfahan and neighbouring cities over the next 25 years" (Murray-Rust and Droogers 2004).

Between 2000 and 2010, an average of 2 BCM of water was used in the Zayandeh Rud catchment for agriculture, industry, drinking water supply and transfers to neighbouring provinces. This was more than the water available from the dam and from groundwater recharge put together. The situation is aggravated by the fact that there have been great fluctuations in water outflow from the Zayandeh Rud reservoir, varying from 533 MCM to 1720 MCM per year (figures provided by the Isfahan Water Authority). Molle et al. (2009) stated that prior to the next flood season, all the water entering the reservoir would be released. This situation of "no significant year-to-year carry-over storage [...] maximizes the production from irrigated agriculture (at the expense of security in supply), and part of the variability in supply is handled by resorting to groundwater". As a result, groundwater levels have decreased considerably and the aquatic habitat has been destabilized.

The different water uses in the catchment lead to diminishing water quantity, but also quality. Research has revealed that on its way from the dam to the salt lake, the river water is used and fed back into the river more than three times. As a result of return flows from agriculture and no or inadequately cleaned urban and industrial effluents, the water shows high concentrations of salt and other pollutants in downstream regions (Salemi et al. 2000). Groundwater resources in several sub-catchments are salinized or at risk to salinization because of deep extraction, so that they can be used only for certain purposes such as cooling water for industry (Iranian Ministry of Energy 2003; Kalbassi 1992). Decreasing water quality, in this respect, aggravates the problem of water quantity.

With almost 90%, the biggest share of surface and ground water resources in the catchment area is being used for irrigation (see Chap. 6 Agriculture). It is in the nature of things that in dry years with little rainfall, water demand for agriculture rises further. Agriculture, thus, has suffered in particular from recent water supply

[2]See https://rsis.ramsar.org/ris/53 (accessed July 27, 2015).

problems. For the distribution of surface water, irrigation systems with multiple distribution structures and channel systems were built in the 1970s and late 90s. Additionally, a lot of ground water is extracted from 45,000 wells for irrigation purposes. Traditionally, farmers irrigate their fields with flood irrigation. Water demand for agriculture has not been met in recent years, not even quantities guaranteed by law. Many *Qanat* systems, sloping underground channels with a series of vertical access shafts, which have been traditionally used for oasis cultivation, had and have to be abandoned due to a disturbed hydrological balance as a result of deep wells. Regardless of that, government and farmers intend to increase agricultural areas and intensify cultivation. Without changing irrigation modes or technologies, this means that water demand will continuously grow.

The Zayandeh Rud also provides water for the second largest industrial area in Iran, with some highly water intense businesses (see Chap. 7 Industry). Since further industrial settlements are planned in the future, the Ministry of Energy's requirement is that the industrial surface water withdrawal must not exceed 200 MCM per year.

Population growth will presumably lead to higher water demands for drinking water purposes in the future (see Chap. 8 Drinking Water). According to forecasts, population in the catchment will rise from around 4.5 million today to more than 5 million in 2020 (according to scenarios developed throughout the project). In case of constant drinking water consumption, the Ministry of Energy estimates an increase in drinking water supply with water resources from the Zayandeh Rud to a total of 400 MCM by 2020. There is a growing responsibility to provide water for other cities with several hundred thousand inhabitants like Yazd, Kashan, Naein, Natanz and Ardestan. Further water transfer projects to Shahrekord, Broojen and Ben are currently under construction.

Water demand of the Gavkhuni Marshland at the river's estuary has not been met for years, and without changes in water management the situation is expected to worsen. Fauna and flora of the salt lake highly depend on inflows from the Zayandeh Rud River. In general, wetlands are very vulnerable systems and highly sensitive to changes in water quality or quantity. In order to preserve this diversity of aquatic and amphibian habitat, a steady, sufficient inflow would be necessary. Current inflow is estimated to be less than 30 MCM per year on average. Additionally, the water that reaches Gavkhuni is of very low quality because of extreme sulphate load and high oxygen depletion (Pourmoghaddas 2006).

All the different ecological conditions and settlement structures described affect the quantity and quality of water resources. With constant yearly renewable water availability of 2 BCM, the deficit between water demand and water availability in the Zayandeh Rud catchment will continue to grow.

3.3 The Role of the Zayandeh Rud in the Region's History

The history of Isfahan and its great reputation is strongly tied to the Zayandeh Rud. Since access to water has always been crucial in the history of civilization, the Zayandeh Rud River played and still plays a vital role in this hot arid zone, as it is the main source of water in the middle of the Iranian plateau. The Zayandeh Rud with its unique characteristics has been widely noted in historical documents as the main source of water in central Iran (Mustawfi 2002; Shafaghi 2002; Mafrukhi 2006) giving life to the land and villages of the provinces of Isfahan, Chahar-Mahal and Bakhtiari and Yazd, thereby being the foundation for an evolving and flourishing civilization on the banks of the river.

According to historical evidence, early human settlements were formed in the Isfahan plain. In Palaeolithic times people were drawn towards the river of Zayandeh Rud as a reliable source of water (Conard et al. 2005) which is underlined by findings of Biglari et al. (2009). They found artefacts and tools in a cave called Qaleh Bozi overlooking the river, the origin of the stones used for tool making were the shores of the river, showing that the river had an impact on almost every part of life in the Palaeolithic age. Furthermore the Isfahan plain was inviting to live in due to good weather, the flow of the Zayandeh Rud and the fertile soil on the banks of the river (Shafaghi 2002). As a further development in human civilization in Iran, the origin of agriculture can be found in the foothills of the Zagros Mountains, this being part of the Fertile Crescent (Riehl et al. 2013). These characteristics make the Zayandeh Rud catchment one of the most important regions in Iran, having always contributed to the history of Iran.

The city of Isfahan goes back to the ancient region called Aspadana. During the empire of the Median dynasty (ca. 675 – 550 BCE), Aspadana was part of the area where the Median tribe Paretaceni resided in an area called Jey today, formerly Gabbay (Schwarz 1969; Diaconof 2011). In the pre-Islamic era, Jey referred to the vast area including the north and south banks of Zayandeh Rud River and the river flowed through the lands of Jey. Today, Jey refers to a region of Isfahan including the lands on the north banks of the river (de Planhol 2006). It is also well known as one of the most historic neighbourhoods in modern Isfahan.

The regions of the Persian and Median empires were united by Cyrus the Great, creating the beginning of the Achaemenide era (ca. 650 – 330 BCE), a prosperous time in terms of cultural, political, religious, social and economic development of Isfahan (Miller 2004). Jey, being located in the intersection of the main roads of Iran, was one of the most important regions in the Achaemenide Empire; consequently the Achaemenide kings resided in this region for better control of the area.

After the invasion and conquest by Alexander the Great, the city of Jey further established its role as a trade city (Hosseini Abari 2008). In the following Sassanid era (ca. 225–650 CE), the Zoroastrian city of Jey and the Jewish city Yahudiya, which was located around three kilometres north of Jey, merged into the city of Isfahan (de Planhol 2006).

Isfahan was home to a number of noble families and a place for the military education of the crown princes. A further indication of the importance of Isfahan during this period is the fact that it was ruled by the crown prince. This is in accordance with Sasanians tradition, which demands an important area to be ruled by the crown prince as training for kingship (Honarfar 2010).

After the capture of Isfahan by the Arabs, marking the beginning of the Islamic era, the region was ruled by the Al-Buyid dynasty (934–1062 CE). During the Al-Buyid era, small villages and cities were integrated into the city of Isfahan and new neighbourhoods were created. This led to further urbanization and prosperity (Hosseini Abari 2008; Omrani Pour et al. 2012) and provided the foundation for further development in the region. The succeeding dynasty of Turkish Seljuqs (1037–1194 CE), in particular its founder Toghril Beg who resided in Isfahan for twelve years, made the city of Isfahan the capital of the empire. Under these circumstances the city grew and flourished (Honarfar 2010). In the twelfth century, the most important aspect of the city in relation to the Zayandeh Rud was the existence of the network of fresh water channels and irrigated gardens in the city of Isfahan (Omrani Pour et al. 2012) which distinguished the city from other Persian cities.

After the fall of the Seljuq dynasty, Isfahan experienced periods of decline until the seventeenth century when it became the Safavid dynasty's capital and the golden years of splendour started. It is said that the Zayandeh Rud and its influence on fertility in Isfahan region was one of the main reasons that Shah Abbas decided to move the capital from Qazvin to Isfahan in 1598 (Baykal 2007). Isfahan experienced several years of urbanization before flourishing during the time of the Safavid dynasty. The distinctive characteristic of Isfahan, including the unique geographical location in central Iran and particularly, the flow of Zayandeh Rud and its surrounding fertile lands, made Isfahan one of the most important regions in central Iran. During the early centuries of Islam, the region of Isfahan included thirty rural areas (Rustag) and more than one thousand small villages. In this period, Isfahan was well-known for its higher quality of agricultural products as one of the main reasons for its economic prosperity (Hosseini Abari 2008).

It was the geographical interaction between the Zayandeh Rud and Isfahan that led to development and urbanization in the Safavid period. Because of the political and economic activities directed by Shah Abbas, the population grew through immigration by migrants from the Caucasus region. This population increase resulted in a pressing need for higher agricultural production so as to provide enough food for the population and the imperial system. Therefore, an efficient distribution system for the water of Zayandeh Rud was needed (Omrani Pour et al. 2012). With this in mind the history of the flourishing years of the city during the Safavid dynasty is inextricably interwoven with the existence of the Zayandeh Rud. There were, for instance, around 100 streams called Madis flowing from the river in the city towards the Jey lands. Madis are man-made channels, the name being adopted from the Mad dynasty which ruled the country three to five centuries BCE. Twelve bridges, some examples of magnificent architecture, were also constructed, with the special features of serving both as bridges and dams (Hosseini Abari 2008,

p. 75; Omrani Pour et al. 2012). The function as dams, and thereby giving an option for active water management, was particularly important as low water resources have always been one of the main problems in Iran impacting all aspect of social and cultural life. Hence, dealing with the water problems was a critical issue for the dynasties and empires. It was specifically more critical in the semi- arid areas in central Iran including Isfahan region. Throughout history in the region of Isfahan the existence of the Zayandeh Rud has been tied to the development of civilization in the area, displaying the relationship between human culture and natural resources.

3.4 Historical Water Distribution Rights: Sheikh Bahaei Water Share Scroll

Due to the essential role that Zayandeh Rud has consistently played in irrigating the fertile lands in the Isfahan plain, the ways of distribution of water have always been a critical issue, sometimes leading to conflicts and tensions among water users in the region.

Regarding the historical texts, the use of Zayandeh Rud water was always based on a specific pattern throughout its history. Since individuals using water of Zayandeh Rud built streams branched from Zayandeh Rud – the Madis – some scholars believe that the basic concept of the water distribution pattern of Zayandeh Rud can be traced back to the Median Empire (Mehryar 1999; Islami 2009). In this sense, the history of developing an efficient water distribution system is also the history of using the water of Zayandeh Rud as the main source of water in central Iran. One important feature of the distribution system has to be the consideration of water shortage and the equal status of all stakeholders in the access to water resources at the same time.

Based on what has been written in historical texts, the history of the Sheikh Bahaei water share scroll goes back to a special water distribution pattern attributed to the famous scientist and scholar Sheikh Bahaei. It was set in the Safavid Dynasty after the reign of Shah Abbas I (1571–1629), under the regency of Shah Ismail III and was later adapted during the Shah Tahmasp II era to the water needs in this time period. It was used as a guideline for water use and water distribution in the Isfahan region (Mehryar 1999). The scroll is a legacy of insight into the historical, cultural and social evolution of agriculture of the Zayandeh Rud basin. Although this historical document was referred to as the Sheikh Bahaei scroll, it is not perfectly clear why this document was attributed to Sheikh Bahaei. The scroll based water distribution pattern has been tried and implemented during centuries, and has adapted itself to the necessities of the region.

The scroll has special peculiarities which have given authentication to its practicality:

1. Its importance and historical value as a world heritage in water distribution within areas with low precipitation and severe water dependence.
2. A historical example for participative cooperation and decision making in the right to water and its distribution, maintenance, utilization and surveillance.
3. Consideration of varying necessities due to different cultivation patterns, plantation sequences and water rights of the Zayandeh Rud.
4. General acceptance of this distribution pattern by all people and observance of its regulations.
5. A successful and dependable pattern for a multilateral cooperation of stakeholders in water utilization of an area with the most sensitive and tangible social concerns.

According to the historical document the water of Zayandeh Rud should be allocated among the Madis mentioned in the scroll. Although the amount of water allocated to each share was not mentioned clearly in the scroll, there were two main criteria to identify the pattern of water distribution which included time and the location of the Madis (Mehryar 1999). In this way, the water of Zayandeh Rud was distributed in terms of 33 shares which have been divided into seven blocks including Lenjan, Alenjan, Marbin, Jey, Baraan, Kararaj and Roudashtein (Islami 2009). The exact distribution is listed in Table 3.1 and will be explained below.

The scroll based water distribution pattern of Sheikh Bahaei has organized the spatial and temporal water distribution to some 405 cities, villages and farmlands. The implementation of the said scroll, with its vast and huge territory, containing so many cities and villages with its social, agricultural diversities, was solely decided by the stakeholders or their representatives. Even the highest ranking organizational person on utilization of the river being called "water head distributor", which today corresponds to the managing director of the Water Board Co., was elected every spring by stakeholders for a period of one year and was introduced to the Government. This structure is depicted in Fig. 3.4.

In order for downstream users to supervise water extractions by upstream users according to the scroll, there was a nominated surveillance board which was selected by downstream water rights owners (Fig. 3.5).

The head of water distribution, supervised by the downstream users, along with the stakeholders advise and regulate the upstream head of distribution and the water staff, this being the executive releasing the water to the users. The water staff consists of the head watchmen and the block stream distributors of the seven blocks, which are the final consuming blocks.

The main pattern of the Sheikh Bahaei water share scroll is based on a crop calendar and percentage of available water. This means that in the first 75 days and in the last four months of a year the water of the Zayandeh Rud is shared among water rights holders. In other words, in this free extraction period, water is extracted by water rights holders commonly. After the first 75 days the scroll based distribution is executed, spanning over a period of 165 days, which starts from June 5th (15th Khordad) lasting until Nov 20th (30th of Aban). The exact distribution basis is listed in Table 3.1. According to the scroll, the Zayandeh Rud water is divided

Table 3.1 Distribution of water among the scroll-based blocks: Roudastein, Kararaj & Baraan, Marbin & Jey, Lenjan & E Lenjan (Source: Islami 2009)

Month	Irrigation days	Block			
		Lenjan & E Lenjan	Marbin & Jey	Kararaj & Baraan	Roudastein
May 22–Jun 21 (Khordad)	16–30				15 days seeding water
Jun 22–Jul 22 (Titr)	1–18	18 days			
	19–30		12 days		
Jul 22–Aug 22 (Mordad)	1–15	9 days			
			6 days		
	16–30	9 days			
			6 days		
Aug 22–Sep. 22 (Shahrivar)	1–15	9 days			
			6 days		
	16–30	9 days			
			6 days		
Sep 22–Oct 22 (Mehr)	1–18	11 days			
			7 days		
	19–28	10 days sowing			
Oct 22–Nov 22 (Alban)	29 Mehr–8 Alban		10 days sowing		
	9–15			7 days sowing	
	16–30				15 days budding water

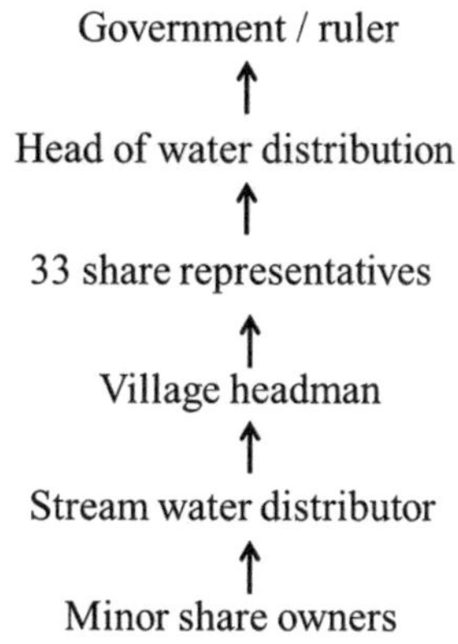

Fig. 3.4 Organizational chart of Zayandeh Rud utilization body elected by all stakeholders within the territory

Government / ruler

↑

Head of water distribution

↑

Head watchman (one person from each water rights owner block)

↑

Stream water distributor

Fig. 3.5 Organizational chart of the surveillance board nominated by down-stream water-head distributor for supervision of water extraction

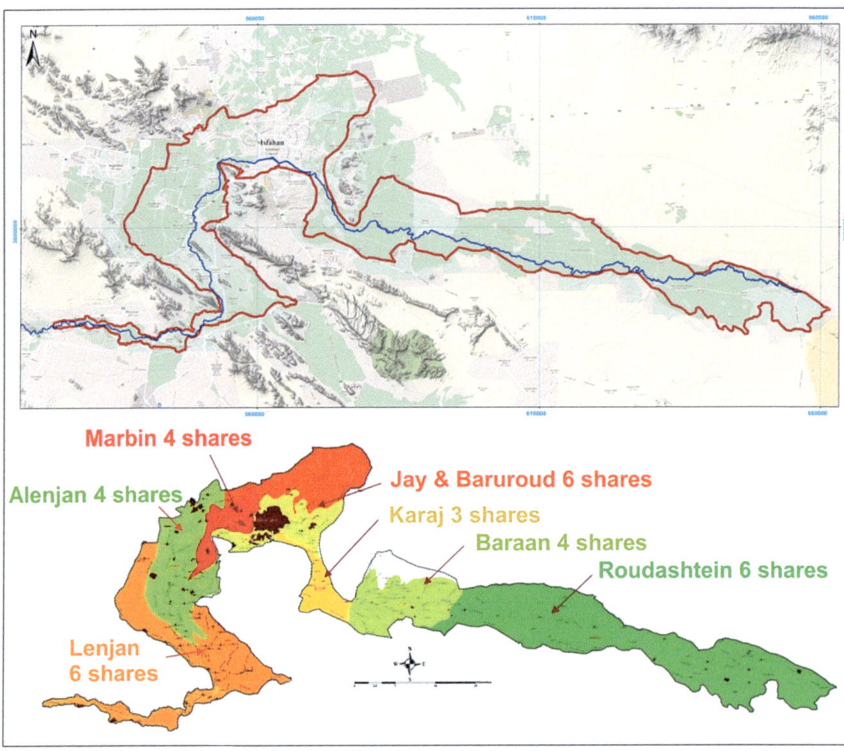

Fig. 3.6 Distribution of water based on the scroll-based in seven right-owner blocks (Source: Isfahan Water Authority)

into 33 major shares, and for each major share, five days of extraction have been allocated. In order to support the principles of the Sheikh Bahaei scroll, the water from the nearby river Kuhrang is transferred through the first Kuhrang tunnel to the Zayandeh Rud. This water is also divided into 33 shares and distributed between the blocks (Fig. 3.6).

Table 3.2 Subdivision of major shares to minor and retailed shares in the Sheikh Bahaei water distribution scroll

	Lenjan	Alenjan	Marbin	Jey-Barzroud	Kararaj	Baraan	Roodasht
↓33 major water shares	6 shares	4 shares	4 shares	6 shares	3 shares	4 shares	6 shares
↓275 minor water shares	113,5 shares	47,5 shares	29 shares	37 shares	12 shares	14 shares	22 shares
3098 retailed water shares	357 shares	315 shares	282 shares	674 shares	387 shares	840 shares	243 shares

Regarding the diversity of cultivation in the upper, median and lower parts of the Zayandeh Rud, within the limits of the scroll, each of the 33 shares is entitled to extract water in a limited time. In this irrigation sequence, special attention has been given to cultivation sequence of rice, kitchen garden, crops and considering the time spans in which plantations are water-sensitive (budding time, flourishing, seeding). Furthermore, the scroll has allotted a water right for Isfahan metropolitan with some priorities.

Table 3.1 shows the irrigation sequence and temporal distribution of water based on the scroll. In the irrigation sequence, each of the scroll blocks can extract water for a certain time span. This irrigation sequence is applicable similarly for each major share and all shares in every block.

In the scroll, there are seven right-owner blocks and 33 shares of water from the Zayandeh Rud. Each of the 33 shares in terms of water extraction is identical and equal to 5 days of extraction. Therefore, due to different river discharges (flow) in different months, the volume of water is changeable in different blocks. In the scroll the major shares are subdivided into minor and retailed shares, with regards to local and regional specifics, the number of villages, traditional networks, ownership of lands and vast farming areas. The major shares are broken into 275 minor shares and 3098 retailed shares. The exact basis of subdivision is listed in Table 3.2. It is noteworthy that although the minor and retailed shares in one area are equal, in different blocks these shares are not equal in terms of time-span extraction and volume.

3.5 Current Water Distribution Issues in the Catchment

While changes in the catchment and general socioeconomic conditions have led to changes in the system of water distribution in Isfahan, what was written in the Sheikh Bahaei scroll still remains to be the main acceptable pattern to distribute water of the Zayandeh Rud among traditional water users, in particular local farmers. The distribution found in the Sheikh Bahaei scroll was adopted by law in 1954 as the main pattern of water distribution in Isfahan including the mentioned water rights holders. This distribution was also referred to by the Law of Equitable Distribution of Water adapted in 1982. However, it has not been clear and is still a subject of ongoing discussions, what amount of water from the Zayandeh Rud should be allocated to traditional water rights holders and how new stakeholders like industrial companies should be considered.

At present the amount of water distributed is based on time, location and particularly the percentage of water available in the river. This means that based on water availability in the river, the water is allocated among the 33 water shares (Mehryar 2000). As a result, in times of water shortage the process of water distribution has always become a critical issue in the region. As mentioned before with regards to changes, like the extension of cultivated lands leading to the development of new irrigation networks, the water distribution system has always been changed consequently.

Among several causes, the extension of cultivated lands and the development of irrigation canals to address the population growth and the establishment of industrial settlements in the region of Isfahan are the main reasons leading to increasing numbers of water users within the catchment. Water managers today face the daunting challenge of securing an increasing water demand with decreasing water resources. Conflicts between the main water users and uses arise. A sustainable reallocation that is adjusted to satisfy all users' demands is the major challenge in the catchment.

References

Baykal H (2007) Vom Perserreich zum Iran, 3000 Jahre Kultur und Geschichte. Konrad Theiss Verlag, Stuttgart

Biglari F, Javeri M, Mashkour M, Yazdi M, Shidrang S, Tengberg M, Taheri K, Darvish J (2009) Test excavations at the Middle Palaeolithic sites of Qaleh Bozi, southwest of central Iran, a preliminary report. In: Otte M, Biglari F, Jaubert J (eds) Iran Palaeolithic, BAR International Series S1968. Archaeopress, Oxford, pp 29–38

Conard NJ, Ghasidian E, Heydari S, Zeidee M (2005) Report on the 2005 survey of the Tübingen-Iranian Stone Age research project in the provinces of Esfahan, Fars and Kohgiluyeh-Boyerahmad. Archaeol Rep 5:9–34

de Planhol X (2006) Isfahan–Historical geography: an overview since ancient times. Encyclopaedia Iranica, Vol. XIII, Fasc. 6, pp. 617–622

Diaconof AM (2011) The history of medians. Elmi Farhangi publication, Tehran

Honarfar L (2010) Isfahan. Elmi Farhangi Publication, Tehran

Hossaini Abari SH (1991) Zayandeh Rud and Surface Water Use. In: Faculty of Literature and Human Science. University of Isfahan, Isfahan. No. 3. (in Farsi)

Hossaini Abari SH (2000) Zayandeh Rud from Its Source to the Mouth. Golha, Esfahan

Hosseini Abari SH (2008) Isfahan rahavard e Zayandeh Rud, cultural organization of Isfahan municipality, Isfahan. (in Farsi)

Iranian Ministry of Energy (2003) Water resources and water demand balance in the Zayandeh Rud catchment. (in Farsi)

Islami M (2009) Zayandeh Rud, Past, present, future (in Persian: Zayandeh Rud, gozashteh, hal, Ayandeh). In: The Conference of Crisis of Zayandeh Rud, cultural organization of Isfahan Municipality. (in Farsi)
Kalbassi M (1992) Salinization of Zayandeh Rud and its changes. In: Wasser und Abwasser. No. 8. (in Farsi)
Mafrukhi (2006) Mahasen Isfahan, cultural organization of Isfahan municipality, Isfahan
Mehryar M (1999) The scroll of the water share of Zayandeh Rud, Farhang Isfahan, vol 12, pp 55–58
Mehryar M (2000) The scroll of the water share of Zayandeh Rud, Farhang, Isfahan, vol 15, pp 26–34
Miller MC (2004) Athens and Persia in the fifth century BC: a study in cultural receptivity. Cambridge University Press, Cambridge UK, p 243
Mohajeri S, Dierich A (2008) Targets for the Iranian wastewater sector. Future prospects and strategic approaches, Report for the World Bank, Berlin
Molle F, Ghazi I, Murray-Rust H (2009) Buying respite: Esfahan and the Zayandeh Rud River basin, Iran. In: Molle F, Wester P (eds) River Basin trajectories: societies, environments and development. Comprehensive assessment of water management in agriculture No. 8. CABI, Wallingford
Murray-Rust H, Droogers P (eds) (2004) Water for the future: linking irrigation and water allocation in the Zayandeh Rud Basin, Iran. International Water Management Institute, Colombo
Mustawfi H (2002) nezhatolgholub, Hadis Emruz, Ghazvin. (in Farsi)
Omrani Pour A, Moradi A, Mohammad FSM (2012) Natural Environment of Zayandeh Rud and the Safavid development of Isfahan. Int J Architec Eng Urban Plan 22(2):114–118
Pourmoghaddas H (2006) Water Quality and Health Issues in the Zayandeh Rud Basin. IWMI Research Project., http://tinyurl.com/ovqlvwc (Accessed 03 Sep 2015)
Riehl S, Zeidi M, Conard NJ (2013) Emergence of agriculture in the foothills of the Zagros mountains of Iran. Science 341(6141):65–67. doi:10.1126/science.1236743
Sahafii A, Sadeghi M (1997): 4th report on vegetation, knowledge about past and present situation, results and future orientation, 2nd edition (in Farsi)
Salemi HR, Mamanpoush A, Miranzadeh M, Akbari M, Torabi M, Toomanian N, Murray-Rust H, Droogers P, Sally H, Gieske A (2000) Water management for sustainable irrigated agriculture in the Zayandeh Rud Basin, Esfahan Province, Iran. IAERI-IWMI research reports No. 1, Karaj, Iran/Colombo
Salemi HR, Murray-Rust H (2002) Water supply and demand forecasting in the Zayandeh Rud basin, Iran. IAERI-IWMI research reports No. 13, Isfahan/Colombo
Sarhadi A, Soltani S (2013) Determination of water requirements of the Gavkhuni wetland, Iran: a hydrological approach. J Arid Environ 98(2013):27–40
Schwarz P (1969) Iran im Mittelalter nach den arabischen Geographen, 9 vols., Leipzig, Zwickau, and Stuttgart, 1896-1935; repr. in one vol., Hildesheim and New York, 1969; see vol. V, pp. 587–90
Shafaghi S (2003) Geography of Esfahan. Isfahan University Publishing, Isfahan
Shafaghi S (2002) The geography of Isfahan. University of Isfahan, Isfahan

Part II
Developing a Common Understanding of IWRM and Its Requirements

Chapter 4
Participation in Water Management in Iran

Lena Horlemann and Parisa Jafari Berenji

4.1 Introduction

The abundance of public attention and debates focusing on the issues and challenges related to public participation in Iran shows how this issue has become central in recent years (Namazi 2000; Afrasiabai 2003; Bradley 2007). Public participation in the process of decision-making and management has become more important in Iranian public debates after the Islamic Revolution in 1979. While the constitution calls for "the participation of the entire people in determining their political, economic, social and cultural destiny" (Chap. 1 Art. 3), at the same time it states that the final decision is taken by the highest clergy. Despite its increasing importance over the last years, there is no common perception of how participation of social actors should influence decisions and management and particularly, how participatory decision-making can be embedded and thus institutionalized in the centralized political and decision making structure of Iran (Bradley 2007). After the election of Khatami in 1997, a time perceived as promoting 'greater freedom' for Iran's citizens, Namazi (2000, p. 13) stated:

> Iran faces daunting challenges in its drive for more participation of all citizens, civil society development and NGO empowerment. The process of political change faces serious obstacles and hurdles. In the absence of the culture and tradition of political participation, the process of change entails ebbs and flows and is far from smooth.

According to participatory management approaches, an effective participatory process should involve interaction between social actors and decision-makers at multiple levels of decision making. Stakeholder participation as a main element for sustainable environmental and water management allows comprehensive and reflexive definitions of problems and incorporation of underestimated points of

L. Horlemann (✉) • P. Jafari Berenji
inter 3 GmbH Institute for Resource Management, Berlin, Germany
e-mail: horlemann@inter3.de

S. Mohajeri, L. Horlemann (eds.), *Reviving the Dying Giant*,
DOI 10.1007/978-3-319-54922-4_4

view (see Global Water Partnership GWP 2000; Mostert 2003; Pahl-Wostl et al. 2007; Kirschke et al. 2016). Based on the participatory management approach, stakeholder participation can take place at various levels. But one of the main ideas is the inclusion of all relevant stakeholders, i.e. decision-makers, social and economic sectors, and affected people. Stakeholder involvement can be applied in the phases of problem definition, data collection, interpretation of results and development of policy options. Furthermore, participation can facilitate the relation between stakeholders and it can contribute to setting the right priorities and making the process of decision making more transparent. In addition, it helps to avoid narrow or partial examinations, and help not to exclude any of the stakeholders (Nasrabadi and Shamsai 2014, p. 793).

Participation as a process in which stakeholders influence policy formulation, strategy design and management (Yercan 2003) needs to be considered in the socio-political context. In this sense, institutional factors such as political structure and legal framework can affect the process of participation. In the context of Iran, participatory process has been conceptualized in a variety of forms implying different meanings and interpretations which are applied in practice in several ways. A major obstacle seems to be the lack of a participatory tradition and persistent governance structures. Namazi (2000, p. 13) states:

> Old legal and procedural forms [. . .] are still in place. Even worse is the negative attitude of senior officials in the executive, judicative and legislative branches that need to be overcome.

The historical background of water management in Iran shows that the traditional way of water management reliant on local participation management of water distribution and allocation has steadily changed since the white revolution and land reforms (Hosseini Abari 2008, p. 112) launched by Mohammad Reza Pahlavi in 1963. It was a reform from above changing the power structure in rural communities and between rural communities and other levels of Iranian society as well. According to Hosseini Abari (2008), this led to a "governmentalisation" of water management meaning that the central government became the main responsible body to provide and manage public needs including water resources, and the role of local peoples' participation was degraded. Furthermore, water was no longer considered simply a need for people's livelihood but an essential resource for economic means like industrial and agricultural production.

Since political and administrative institutions expanded and government became more centralized, particularly after the 1960s, the government has played a central role in planning and budgeting (Ardakanian 2005), and consequently natural resources have become a property of the state (called nationalization of natural resources). But, since the users of the natural resources are people of regions in which the resources are located, the government has not been able to fully control the use of resources in practice (Bagherian et al. 2009, p. 429).

While the Islamic Revolution of 1979 led to changes in the political and social system through officially supporting and encouraging public participation, a centralized and top-down management structure has still remained, which is also dominant in water resource management, and evidence shows it has caused several issues and problems in the context of water management and policy design.

4.2 Current Water Management Issues

Evidence shows that one of the main problems affecting the process of sustainable participatory management in Iran is that because of the centralized political structure, there is no efficient cooperation between different sectors and government bodies, which has resulted in a lack of common understanding about the problems, issues and capacities and also conflicts between and amongst decision makers and water users (see Mohajeri et al. 2016). Consequently, in these conditions, most of the important decisions are taken by the highest rank of authorities which in most cases are not accepted and supported by other stakeholders.

Cooperation between the main governmental organizations including the Ministry of Energy and its subordinated organizations, other ministries of water using-sectors, the provincial agricultural, industrial or environmental organizations, is very low and so the decisions are very selective. As a consequence, they can barely be harmonized with decisions and policies taken in other sectors or by other organizations. In addition, decisions and policies are not able to represent the interests of all social groups who could be affected, since they are not part of a participatory integrative process of decision making and management.

Regarding Iranian water management, improving structural aspects of water resource management has been highly considered in regional and national development plans and policies. In this regard, supporting public participation is one of the main orientations of the plans in order to improve water user systems and increase economic efficiency of water (Tahbaz Salehi et al. 2010). This kind of orientation not only indicates the challenges in Iran's water sector, but also acknowledges the importance of the incorporation of sectors including all levels of water organization, stakeholders and water users, and also between private and public sectors in the process of water resource management. It therefore implies that without public participation it is hard to reach the main goals of development plans related to the water sector.

As evidence shows, where participation of water users was ignored, most of the water sector related plans did not achieve complete success (Shahroudi and Chizari 2007). For instance, since water scarcity is the most limiting factor in the agricultural sector in Iran, the issue of water users' participation in irrigation water management has become more important, as the majority of the plans should be implemented by farmers. Consequently, their acceptance and support is essential. Therefore, it is necessary to consider farmers' participation in the processes of policy design including addressing common problems, capacity building, and policy implementation.

Considerable attempts have been made to use local capacities in operating water-related plans. One of the most important, for instance, is to commit operation and maintenance of irrigation and drainage networks to farmers and to support them by providing financial facilities (Ardakanian 2005). In spite of challenges such as providing economic resources for these types of participatory plans, as experience shows, they could be more successful compared to the plans which have only been managed by governmental sectors.

While important efforts such as providing a legal framework have been made to include public involvement in the process of decision making in recent years, participatory management has not really institutionalized yet in the centralized political structure of Iran. In the water sector, for instance, the aspect of participatory implementation of projects and policies is more important than the participatory policy formulation or decision-making process. Sectorial interests and positions continue to play a main role in these processes. IWRM and participatory water management principles would still need to integrate different concerns and interests from all affected sectors and social groups to identify a common understanding about the problems and challenges.

One of the challenges is the lack of skills required to implement the process of participatory water management in Iran. Thus capacity development is necessary (Nasrabadi and Shamsai 2014) to enhance skills as well as professional and scientific-based knowledge which can be applied in the process of decision-making. Another issue is providing transparency in the communication of participatory process aims, methods and phases among all stakeholders, social and interest groups which is another weakness in Iranian water management. There is a need to enhance the co-learning process of building relations and legitimating decisions (Nasrabadi and Shamsai 2014). It should involve all levels of participation from local level to regional and national levels of decision making and policy designing.

4.3 Participatory Definition of Water Management Challenges

One of the aspects of participation in water management is to involve all relevant stakeholders in the process of definition of water management challenges. Involving stakeholders is an important step to ensure that water management decisions and plans take into consideration all local needs, experiences and interests (Stanghellini 2010). Participatory defining of problems and challenges legitimate decisions made to address these problems and challenges in water management. In a legitimate decision, all stakeholders feel that their input, concerns and expectations have influenced the decision, or they believe that the process of decision-making has been made through a fair, transparent and open process. Such decisions are expected to evoke less resistance and therefore to be implemented more successfully (Carr 2015, p. 397).

In the process of the IWRM Zayandeh Rud project, and with the aim of providing a participatory basis for defining water management challenges from

Fig. 4.1 Stakeholder involvement in the IWRM Isfahan project

the viewpoint of stakeholders, an interactive, participatory workshop was conducted in 2012, involving all relevant stakeholders from national and local government, water, agricultural, industrial and environmental sectors, and academia (see Fig. 4.1).

The development of the workshop methodology had to deal with some challenges: The first challenge was the hierarchical system of decision-making which leaves little space for participation or negotiation across hierarchical levels. And second, the form of decision-making: especially in years of water shortage, decisions over water distribution have been taken on an *ad hoc* basis, in an attempt to balance acute water demands. These decisions, however, have not been based on sound data or a long-term management plan. This has led to severe inter-sectoral and regional (between provinces) conflicts of interest, particularly in the Zayandeh Rud catchment area. Due to the lack of transparent decision-making, feelings of unfair distribution or preferential treatment of individual sectors or regions, and since a lack of water mainly puts the livelihood of farmers at risk, these conflicts can be quite emotional.

These challenges could be resolved by addressing the problems in an open way. First, the problem of hierarchical thinking was discussed with the respective authorities and senior participants. Second, three small discussion groups were formed and participants were systematically chosen from different sectors, hierarchical levels and academia. The discussion groups were then chaired by an independent, unbiased person.

The development and implementation of participatory methods was a time-consuming but worthwhile activity since it built an atmosphere of trust and willingness to cooperate between the stakeholders and partners. Moreover, it revealed that the stakeholders are well aware of the main reasons behind water management problems and water stress.

During the workshop, the participants were divided into small working groups to discuss and answer the questions of present and future problems and challenges of water management in the Zayandeh Rud catchment. The answers of each group were then presented to the other working groups for discussion.

4.3.1 Joint Problem Definition

The participants defined four issues as the major water management problems in the catchment:

- Lack of integrated management;
- Lack of data (quantitative and qualitative aspects);
- Water resource shortage;
- Drop of water resource quality.

The lack of integrated management was identified by the stakeholders as the biggest water management challenge. According to their views, the main reasons are sector oriented water management, growing conflicts among different sectors, and the lack of cooperation and mutual understanding between stakeholders for solving problems of the catchment. These issues, again, were observed as consequences of increasing mistrust among stakeholders, and sectoral interests pushed by different stakeholders regarding water resources.

The lack of scientific basis for decision making is another issue that was mentioned by the workshop participants as an obstacle to water management. Considering the issue of the lack of data, the lack of integration of data available in different sectors of the region was mentioned as the main challenge for water management. According to the workshop participants the lack of integrated, updated and scientifically gathered data had a negative impact on the process of decision making and development of integrated water management.

Another challenge stated by participants of the workshop was water resources shortages. The measures taken to date, like water transfers from neighbouring provinces of Chaharmahal and Bakhtiari to the Zayandeh Rud, would not be able to alleviate the problem by themselves and might even exacerbate resource conflicts. On the other hand the stakeholders believed that the inter-basin water transfer would increase in future, due to water consumption growth. Growing water consumption leading to overexploitation of water resources was observed by the stakeholders as a great challenge for water management.

The drop in water resource quality is another water management challenges that was identified by the stakeholders. They made a link between the water quantity and quality in the catchment. In this sense, when the water quantity drops, water resources quality is also negatively affected. They emphasized that both groundwater and surface water had been impacted by pollutant substances which could be an impeding factor for optimal water use. This could also threaten the environment.

Besides the aforementioned issues, during the workshop the stakeholders shared their concerns about the future of water resources and the growth of threatening factors and a worsening of the current situation. Some of the stakeholders were concerned about the negative impacts of the overexploitation of water resources and uncontrolled discharge on the future existence of the Zayandeh Rud river as the major water resource in the Central Plateau of Iran. They believed that the loss of ground and surface water resources could largely affect agricultural sectors and raise local and regional conflicts. In addition, the region would face a severe challenge of losing its strategic industries such as the steel and cement industry. They also anticipated that a loss of clean drinking water would occur due to the increase of harmful ingredients, and could become a threat to peoples' health.

4.4 Participatory Development of a Decision Support System

A decision support system (DSS) is supposed to support scientifically sound, technically robust and unbiased judgments and water management decisions that aim at balancing all water users' current and future interests. This usually means to overcome inter-sectoral conflicts of interests towards water resources. With regards to Iran, or the Zayandeh Rud catchment, these conflicts do not only emerge among individual water users or water user groups, but particularly between their official representatives in respective ministries and other public authorities. The results are an atmosphere of mistrust and a lack of coordination and cooperation. Until today, the Ministry of Energy takes major decisions about water resources management, leading to mostly technological solutions that are rarely agreed on with other key ministries or institutions (Mohajeri et al. 2009), like dam building or water transfers between provinces. Not least as a consequence of these unilateral acts, there were no coherent data or statistics available to be fed into the Water Management Tool (WMT) which was developed in the first stage of the project as a basis of the DSS. The WMT combines the simulation results of three models (MIKE Basin which depicts anthropological impacts on water resources in the catchment area, the groundwater model FEFLOW, and the hydrological model SWAT) and calculates the amount of available water and the supply for each individual user. The WMT will be further developed to a DSS and implemented in the next project stage. Giupponi and Sgobbi (2013) show that for successful DSS development and implementation, not only coherent data are required, but – in a first step – approaches that foster consultation and negotiation among decision-makers.

Experiences with water management tools/DSS have shown that even a careful and practice-oriented development of a model does not guarantee that decision makers will actually use and further develop the model after its implementation (see for example Jao 2011). In their comparison of different projects that have developed a DSS as a steering tool for an IWRM process, Giupponi and Sgobbi (2013, p. 812) found that "the quality of the tools *per se* cannot guarantee the quality of the process". Based on the opinion that acceptance and ownership of the WMT provide the basis for its successful implementation (Serrat-Capdevila et al. 2011), methods for stakeholder involvement were integrated into the project, and locally adjusted means and methodologies for implementing the tool were already being assessed during its development.

In general, the implementation of a decision support system puts previous forms of decision making into question. In Iran, hierarchical thinking prevails and the Ministry of Energy has the final say in water management decisions. The participative development, maintenance and use of WMT/DSS meant negotiating classical working methods and principles of decision making (Ghanavizchian and Mohajeri 2013).

A major challenge of the project was to identify and harmonize the different interests and expectations of the decision makers towards the WMT. Only if the future users see the benefit and their demands are reflected in the tool, will they support its development and implementation. Even if there had been some experience with models in certain areas, there was still uncertainty regarding WMT's essential functions and exact application.

4.4.1 Workshop on Joint WMT Development

For the purpose of clarification, another participative, culturally adapted workshop session was facilitated, following the same methodology as the first workshop on water management challenges. The results of this interactive workshop were again presented in various rounds to different stakeholders. This led to a fruitful discussion within the region about the establishment of new, necessary organizational units which are supposed to manage the IWRM process in the future.

The aim of the workshop was to clarify three main issues regarding the WMT:

- Advantages and expectations of the WMT,
- The issue of data collection, coordination and validation,
- The question of WMT updating and availability.

With regards to the assumed advantages of the WMT and the stakeholders' expectations of the tool, two main points were mentioned. First, stakeholders expect that the prediction and identification of their decisions' consequences will be improved. Second, this will help them to optimize their decisions. Since the tool is fed with scientific as well as socio-economic data, it is capable of analysing the impact of certain water allocation measures on water rights. While the tool is able to visualize how and where decisions may lead to changes in the catchment, it is also helpful in raising awareness of the different facets of water management among the stakeholders. Furthermore, it can assist in taking decisions about new technologies or the location of new industries. Eventually, the WMT may lead to a decrease in social, regional and sectoral conflicts about water resources in the region.

Regarding the question of who should be responsible for data collection and coordination, some critical points have to be addressed. First, up to now data are collected within individual sectors and there is no culture of sharing data. Second, in this atmosphere of mutual mistrust the stakeholders have to accept the actual data that are fed into the WMT. Two proposals were discussed in this regard. The first proposal suggested that an independent committee consisting of experts of the respective regional organizations or sectors should be in charge of collecting the data. Being independent, the committee should at the same time be autonomous enough to be capable of collecting the required data, and it should have the actual mandate to claim due data from defaulting stakeholders. The second proposal suggested that a professional entity, i.e. the Isfahan Regional Water Company, should be responsible for data collection and coordination.

However, the collection and management of data does not only require a capable and acceptable organization. For providing valid data, standards for the measurements and for the data themselves have to be set. This may also require the introduction of new technologies and data collection techniques. Moreover, it was stated that questions of capacity building, adjusted legislation, feedback mechanisms and financing have to be further elaborated on.

The last question that was discussed in the working groups was about the responsible entity for WMT updating and its further development. New (social, environmental, political) trends and developments in the catchment have to be detected and translated into valuable data. The WMT has to be further developed accordingly. Here, three possible organizational solutions were discussed as well: transferring the tasks to a commission, an independent company or consultant or to the Isfahan Regional Water Company.

Some of the workshop participants proposed the formation of a commission consisting of representatives of the important regional decision-makers and users of the WMT as a DSS. This commission would have the task of obtaining from relevant sectors the necessary data, information and proposals for further WMT development as well as managing and monitoring their implementation. They expressed their hope that the formation of a commission would increase confidence in WMT. Transparency regarding the data and information included in WMT also appeared to be an important issue that would in turn increase trust in WMT. The joint decision and development of WMT would at the same time increase the probability and willingness of regional actors using WMT in their decision-making process.

The next proposal concerned the assignment of an independent company or consultant from the private sector for updating WMT. Regional decision makers and users of WMT should mutually decide upon the selection of this company to ensure a successful collaboration with the individual sectors for data and information procurement. The ideas proposed by the experts for an independent, privately-owned company ranged from regional to foreign companies. The role of a foreign company would, however, be limited to monitoring and training an Iranian company commissioned to obtaining the data and information necessary for updating WMT from the relevant sectors, of drawing up proposals for the further development of WMT and of taking responsibility for the their implementation in WMT. In commissioning an independent, privately-owned company, the hope was that the updating and further development of WMT would be exclusively factual and objective, and in this sense uninfluenced by regional decision-makers' particular interests.

Some of the workshop participants propose transferring the tasks to the Regional Water Company. This proposal had been a consequence of the fact that, as one of the most important regional actors, the Water Board Co. bears the responsibility for the management of water resources and is in possession of more information and data concerning regional water resources than actors in other sectors. In contrast to the two previous proposals, this proposal focused on the responsibility for and knowledge of water resources and not on the independence of the institution to be commissioned. Using this approach could perhaps lead to an efficient form of updating and use of WMT; on the other hand, the acceptance of WMT by other sectors will presumably suffer.

4.4.2 Decision on Possible Organizations for WMT Application

The final decision will also depend on the question of which organization is most likely to be trusted by all parties, and which is regarded as being most capable of balancing all interests. Regardless of the decision how the updating and availability of WMT in Isfahan will be regulated, some of the experts proposed involving the Ministry of Energy in the process. The ideas for including its participation range from organizational support to acting as the monitoring institution and the final legal authority in water management. Calling for the involvement of the ministry reflects the fact that the catchment area of the Zayandeh Rud extends to three neighbouring provinces (Chaharmahal and Bakhtiari, Yazd and Isfahan). Involving the Ministry of Energy as the legally responsible institution over and above provincial boundaries, it was argued, would allow for the establishment of an integrated form of water resource management using WMT as well as the acceptance of WMT as a decision-making tool at the national level.

In 2014, a Coordinating Council for Integrated Management of the Zayandeh Rud Basin, the first river basin organization (RBO) in Iran, was established. It is headed by the Ministry of Energy and the governors of the three provinces Chaharmahal and Bakhtiari, Yazd, and Isfahan, the Deputy Ministers of Energy, Agriculture and Industry, Mines and Trade. The representatives of the agricultural unions in the three provinces are its board members (Supreme Water Council, minutes of the 10th meeting 2013). The RBO is supposed to improve the collaboration and coordination of the main stakeholders of the different sectors and provinces. The current vision is that the RBO should be the organization that uses the DSS in the future. The final decision will presumably be taken within the second project phase which started early 2015.

References

Afrasiabai KL (2003) The environmental movement in Iran: perspectives from below and above. Middle East J 57(3):432–448

Ardakanian R (2005) Overview of Water Management in Iran. In: Water conservation, reuse and recycling; proceedings of an Iranian-American workshop. The National Academies Press, Washington DC

Bagherian R, Bahaman AS, Asnarulkhadi AS, Shamsuddin A (2009) Asocial exchange approach to people's participation in watershed management programs in Iran. Eur J Sci Res 34 (3):428–441

Bradley M (2007) Political Islam, political institutions and civil society in Iran: a literature review. Middle East Good Governance Fund (MEGGF), International Development Research Centre (IDRC), Ottawa

Carr G (2015) Public participation in river basin management-an introduction. WIREs Water 2:393–405

Giupponi C, Sgobbi A (2013) Decision support systems for water resources management in developing countries: learning from experiences in Africa. Water 5(2):798–818

Global Water Partnership (GWP) (2000) Integrated water resources management. TAC Background Papers No. 4, Stockholm

Hosseini Abari SH (2008) Isfahan as a result of Zayandeh Rud, cultural organization of Isfahan municipality, Isfahan. (in Farsi)

Jao C (ed) (2011) Efficient decision support systems-practice and challenges from current to future. InTech, Rijeka. doi:10.5772/682

Kirschke S, Horlemann L, Brenda M, Deffner J, Jokisch A, Mohajeri S, Onigkeit J (2016) Benefits and barriers of participation: experiences of applied research projects in integrated water resources management. In: Borchardt D, Bogardi JJ, Ibisch RB (eds) Integrated water resources management: concept, research and implementation. Springer International Publishing, Switzerland

Mohajeri S, Horlemann L, Sklorz S, Kaltofen M, Ghanavizchian S, Nuñez von Voigt T (2016) Integrated water resource management in Isfahan: the Zayandeh Rud catchment. In: Borchardt D, Bogardi JJ, Ibisch RB (eds) Integrated water resources management: concept, research and implementation. Springer International Publishing, Switzerland

Mohajeri S, Hempel L, Dierich A (2009) Analysis of the current organizational structure and legal framework of the Iranian water sector. Identification of strengths, weaknesses and challenges for future development, Report for the World Bank

Mostert E (2003) The challenge of public participation. Water Policy 5:179–197

Ghanavizchian S, Mohajeri S (2013) Coping with cultures-conducting participation projects in Iran. In: Schön S, Mohajeri S, Dierkes M (eds) Machen Kläranlagen glücklich? Ein Panorama grenzüberschreitender Infrastrukturforschung. inter 3, Berlin, pp 108–115

Namazi MB (2000) Non-governmental organizations in the Islamic Republic of Iran: a situation analysis, UNDP-Iran technical papers 1, Tehran

Nasrabadi AT, Shamsai A (2014) Participatory developing strategies for water management in South Khorasan province, Iran. Water Resour 41:792–799

Pahl-Wostl C, Craps M, Dewulf A et al (2007) Social learning and water resources management. Ecol Soc 12(5)

Serrat-Capdevila A, Valdes JB, Gupta HV (2011) Decision support systems in water resources planning and management: stakeholder participation and the sustainable path to science-based decision making. In: Jao C (ed) Efficient decision support systems-practice and challenges from current to future. InTech, Rijeka. doi:10.5772/16897
Shahroudi AA, Chizari M (2007) Farmers' participation in irrigation water management in northeast Iran. In: The 4th Asian Regional Conference & 10th international seminar on Participatory Irrigation Management, Tehran, Iran
Stanghellini PS (2010) Stakeholder involvement in water management: the role of the stakeholder analysis within participatory processes. Water Policy 12(5)
Tahbaz Salehi N, Koupahi M, Nazari MR (2010) The investigation of participatory irrigation management in Iran, case study of cooperative water users in Tajan. J Econ Agric Develop 24 (2):205–216. (in Farsi)
Yercan M (2003) Management turning-over and participatory management of irrigation schemes: a case study of the Gediz river Basin in Turkey. Agric Water Manag 62:205–214

Chapter 5
Towards a Sustainable Advanced Training Sector in Iran: Analysis of Main Obstacles in the Field of Capacity Development

Lena Horlemann

5.1 Introduction

In order to cope with current and future problems of increasing water stress and adequate handling of high investments in water infrastructure, the water sector needs well prepared and educated staff in its water industry and agencies (Ibisch et al. 2016). This also holds true for Iran's water sector (Schön and Mohajeri 2008; Mohajeri and Nuñez von Voigt 2011). The current picture reveals that in general the educational system provides a solid basis in the field of water management, with various organizations at different administrative levels and a distinct (basic) academic and non-academic education. Training centres have permanent and freelance trainers, provide ample facilities and studios.

However, practical and advanced training misses to communicate how to identify and address actual water management problems in day-to-day business. This seems to be the case particularly for governmental training facilities that are the main providers for advanced training in Iran's water sector. There is, though, a political will to adjust the educational system to the changing conditions in the water sector.

Hence, the key questions are: What are the main obstacles that hamper the development of a sustainable advanced training sector in Iran? And how can these obstacles be addressed? The analysis of these obstacles has incorporated different perspectives from the Ministry of Energy, the National Water and Wastewater Engineering Company (NWWEC), regional Water and Wastewater Companies (WWC) as well as Regional Water Companies (RWC) as clients, and training providers (for the institutional structure of water management in Iran see Chapter 2). It deals with obstacles that relate to the current organization of advanced training (rules and

---

L. Horlemann (✉)
inter 3 GmbH Institute for Resource Management, Berlin, Germany
e-mail: horlemann@inter3.de

S. Mohajeri, L. Horlemann (eds.), *Reviving the Dying Giant*,
DOI 10.1007/978-3-319-54922-4_5

responsibilities) as well as operational obstacles (application and non-application of rules). In this way, it also compares normative approaches and operational business.

This is not meant as an exhaustive list, but as a compilation of information and opinions that have been identified by the stakeholders as main obstacles for the water sector.

5.2 Method

The results comprise our own expertise and experiences in the international and Iranian water and educational sectors, supplemented with literature and document research. Interviews with Iranian and German professionals add to this body of knowledge. Two workshops were conducted in September 2015 and March 2016 at the National Water and Wastewater Engineering Co. (NWWEC) with participants from the Ministry of Energy, NWWEC, WWCs and training institutes all over Iran.

The data were evaluated and the results compiled with the help of a constellation analysis, which allows for analysing and visualizing the interaction of different actors and factors (Bruns et al. 2007; Schön et al. 2004, 2007). The fact that the constellation analysis takes into account not only actors and institutions but also technologies, resources, and legal and economic factors ensures that the obstacles analysis is conducted in the required complexity. Visualizing the results helps to place the actors involved within the overall advanced training system constellation.

5.3 Initial Situation and Goals Set by the Iranian Government

Training programs for staff of the Iranian Ministry of Energy as well as its subordinated companies and agencies (National Water and Wastewater Engineering Company, Water and Wastewater Companies, Iranian Water Resources Management Company and Regional Water Companies) are centrally organized by the Ministry of Energy as the main institution responsible for water issues (see Fig. 5.1). For this purpose, the ministry has five divisions that are responsible for the following fields:

- Policy making for training
- Guidelines and standards for training
- Management of training, demand assessment
- Execution of training in and for companies
- Service and infrastructure providers

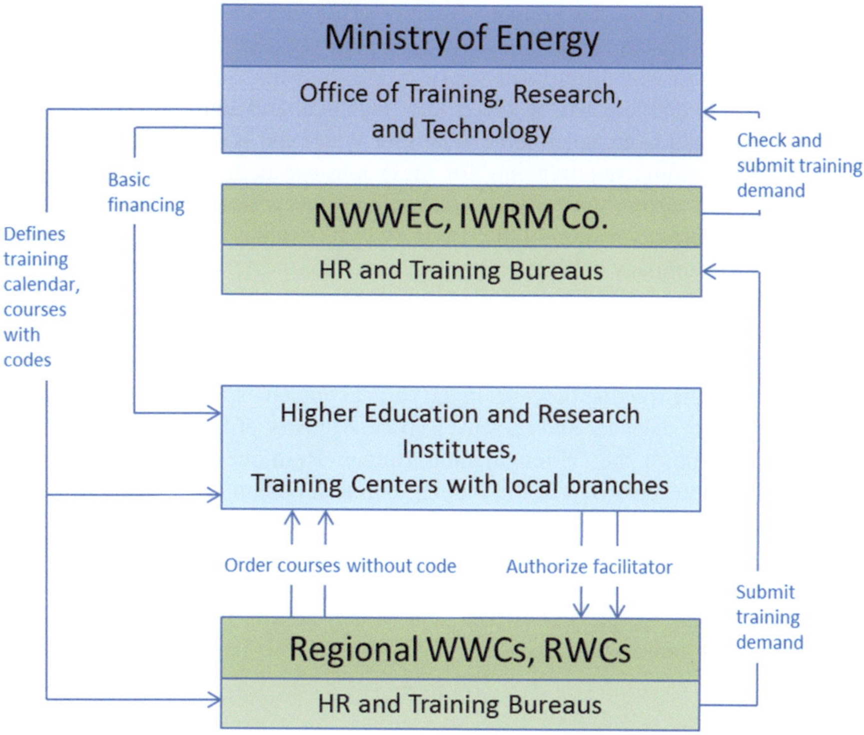

Fig. 5.1 Training system of the Iranian water sector. Own compilation

The ministry's subordinated organizations have human resources and training bureaus that see to training demand assessment and administrative questions like contracts and agreements. Design, management and organization of the training programs are legally based on a directive by the Ministry of Energy.

Around 200,000 people work either for the ministry or its subordinated companies. Of these, about 53,000 work in the water sector. In the years 2011/12 around 24,000 people were employed in Iran's 66 Water and Wastewater Companies, 3000 of them in rural areas and 21,000 in urban areas. 1300 of them held a master's degree or above, 15,900 a university entrance diploma or a bachelor's degree, 6800 were skilled workers).

5.3.1 Demand Assessment and Training Calendar

Due to a lack of interest in advanced training programs, the 30 year old training system was reformed 5 years ago. The aim was to create demand-responsive training programs in order to increase the staff's motivation for taking part in the training. One of the major changes involved the introduction of a bottom-up demand assessment by the human resources (HR) and training bureaus of the ministry's subordinated companies and agencies as opposed to the former process where the demand was determined solely by the ministry. Subsequently, 212 different jobs were defined, 102 of which were technical and administrative jobs for the Regional Water Companies and 110 jobs for Water and Wastewater Companies.

Generally, training requirements for these jobs are defined based on the job descriptions and current standards and technologies. New demands are determined based on the internal situation of the company or its strategy. The HR and training bureau of each company collects training demands for the next year from the heads of department. Employees select their training demands from the current training calendar. This mix of required old and new courses is sent to the parent organization (NWWEC for Water and Wastewater Companies, IWRM Co. for Regional Water Companies) by the training bureaus. In the parent companies, task groups check the proposed training calendar and forward it to the Ministry of Energy for approval.

For this purpose, the "Strengthening Human Resources" expert committee consisting of representatives of the training bureaus, human resources department and external experts come together. Eventually, the ministry approves the training calendar ("comprehensive system of training") and also determines the fees for the individual courses. New courses in the training calendar are given an official code. The national training calendar is then sent to the training institutes that have to prepare for new and existing courses.

5.3.2 Tasks of the Training Institutes and Centres

Training programs are usually[1] facilitated by certified, governmental training institutes or centres.[2] These institutions offer long-term training[3] and degree programs as well as short-term training which until now has only been used by employees of the ministry and their subordinated water authorities and companies.[4] There are 22 such training centres in Iran. They provide training in the fields

[1]There are also independent training centres in Iran, but around 90% of training is done by governmental training institutes and centres.

[2]Training institutes are the regional head offices of training centres and their branches. For convenience, we will only call them training centres in the document.

[3]Long-term studies focus on practical topics like O&M of waste water treatment plants, studies which are not usually offered at normal universities.

[4]With regards to subcontractors, a new regulation has been introduced and from the end of 2016 on, their staff has to do advanced training as well. For the demand assessment of these people, the Technical and Vocational Training Organization of the Labour Ministry and WWC managers defined the training demand according to job descriptions and the international ISO standard. Certificates have to be acquired by 2017.

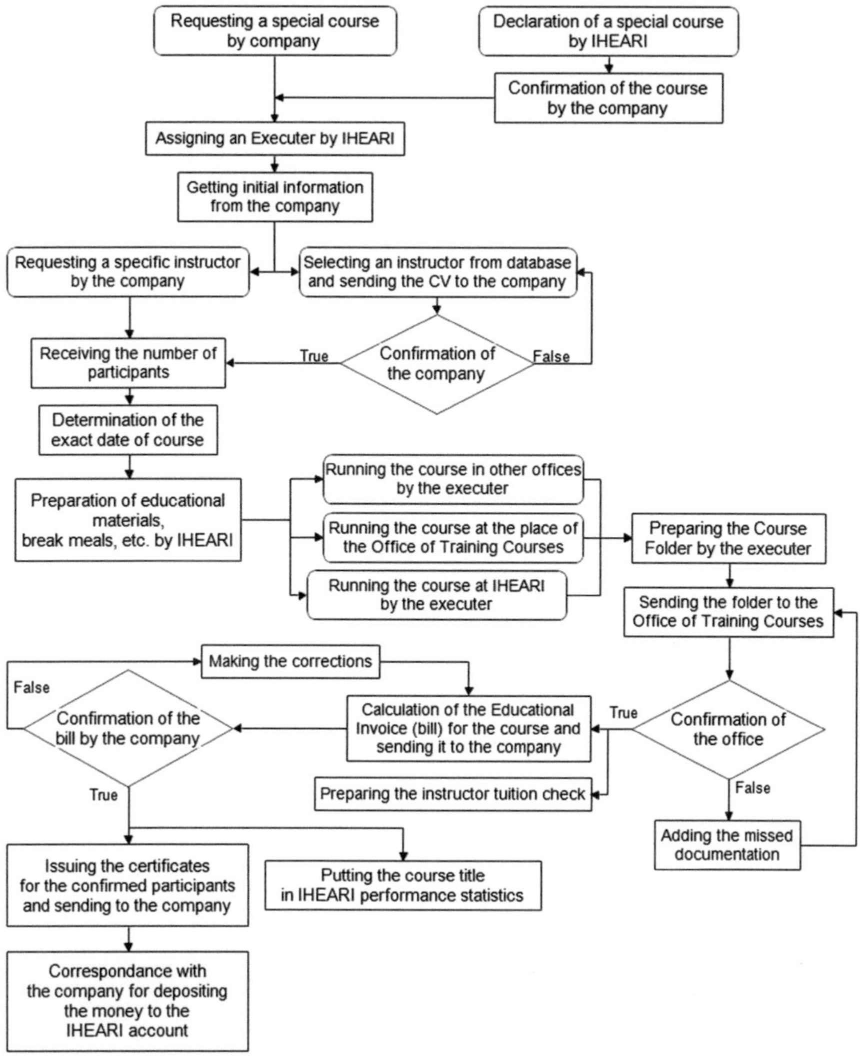

Fig. 5.2 Training Organization Procedure (source: IHEARI)

relevant for the ministry, i.e. water and energy—currently around 1200 courses.[5] Their training also focuses on topics relevant for the respective region, like power plants in the Tabriz training centre.

[5]In the Isfahan Higher Education and Research Institute (IHEARI), for example, about 50% of the courses are conducted by internal staff; the remaining 50% are conducted by external trainers from companies and universities.

In general, there exist three types of advanced training for staff in the water sector: (1) introduction courses which are held at the beginning of employment. The aim of these training courses is to introduce people to the organization and understand its laws and regulations; (2) training that deals with issues like occupational health and safety; (3) job-related training accustomed to people's tasks and expertise to develop further competences for their organizational position (technical, administrative or management).

Moreover, training programs exist for three levels of education: (1) the top educational level, i.e. employees holding a master's degree and above, (2) medium educational level, i.e. (advanced) university entrance diploma, and (3) the bottom level, i.e. workers. At the end of each course, participants have to pass a test before they receive their certificate.

Training centres can, apart from preparing the courses for the national training calendar, act as independent service providers and offer courses directly to companies. These might be courses that were suggested by participants or other experts. Vice versa, companies can order specific training according to their demands at the training centres, which are not part of the official training calendar.

Once a course has been booked by a company, the training centre hires an "executor" who is responsible for organizing the course, communication and facilitation. He is also responsible for choosing and suggesting an adequate trainer. The specific procedure is shown in Fig. 5.2.

5.3.3 Financing of Training

Training fees are paid by the clients, the Water and Wastewater Companies and Regional Water Companies. For this purpose, they provide around 1% of their annual budget. The exact budget provided for training is calculated by the companies' and authorities and submitted to the ministry for final approval.

Budget comes from different sources. A part of the basic financing of the governmental training centres—for facilities, equipment, managers and trainers—is paid by the Ministry of Finance. Additional costs, administrative and service staff, have to be financed through course fees by the training centres.

5.4 Obstacles

What factors mainly constrained the shaping and implementing of a sustainable training system in Iran? This is the key question of the obstacles analysis. Based on the information gathered the following part presents the factors identified and their interrelation and interdependencies. It describes obstacles resting upon (1) the greater, superordinate system, (2) the operational or company level, and (3) the level of the training centres.

This analysis is not exhaustive, but presents the main obstacles identified at the current stage of research.

5.4.1 Obstacles Based on the Superordinate System

There are obstacles in the training system that can be traced back to the overall governance system and business culture (see Fig. 5.3). This includes centralized governance structures, the absence of nation or basin wide water management plans as well as a lack of job prestige, especially in the wastewater sector.

5.4.1.1 Centralism Versus Local Demands

In Iran, the Management and Planning Organization is the central authority for 5-Year-Development-Plans, the ministries' budget as well as the employment of staff for other governmental authorities. The Ministry of Energy is the central body for water management issues. It is superior to the National Water and Wastewater Engineering Co. (the parent company of the Water and Wastewater Companies), the Iranian Water Resources Management Co. (the parent company of the Regional Water Companies) as well as to the governmental training centres. As a result, central government has great influence on the design and procedure of the training system and management of the whole water sector. Conversely, the regional water authorities (WWCs and RWCs) as well as governmental training centres are highly dependent on the superior organizations with regards to planning and organization of training. This situation reflects the general challenges of local or regional organizations in the centralistic governance and state-directed economy of Iran. The current system is regarded as not being flexible enough to respond to or even stimulate changes in the water sector. It seems that decisions taken in the capital are not close enough to the companies' day-to-day business demands, and the inflexible educational system cannot react fast enough to changing demands.

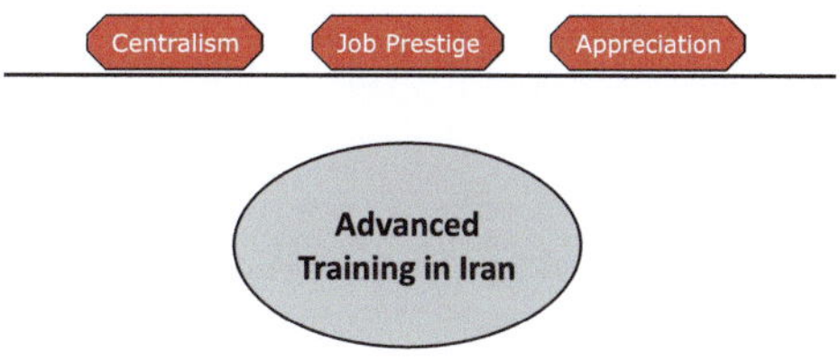

Fig. 5.3 Obstacles at the superordinate level

5.4.1.2 Lack of Job Prestige

Sustainable water management is a critical pillar of the wellbeing of a country and its inhabitants. This is even more the case in countries like Iran with scarce water resources. However, particularly in the wastewater sector, the job prestige is not distinct. Employees of wastewater companies do not consider themselves as environmental protectors or care-takers of human welfare, a phenomenon that can be observed in many countries. This is regarded as a reason for a lack of motivation to participate in training for the purpose of improving WWC operation. In the long run, motivation and reputation need to be increased through a long lasting educational process for society as a whole.

5.4.1.3 Lack of a Water Management Master Plan

A national or basin-wide water management plan usually sets the management goals for the following years. Based on this plan, management activities and measures are planned by the respective government authorities. Hence, a management plan would serve as a basis for the definition of future demands in expert knowledge and training. The lack of an overall water management plan or basin management plans for Iran therefore hampers the development of a demand-responsive national training calendar. Such a plan could, however, serve as a basis for the "Strengthening Human Resources" committee that prepares the national training calendar to take decisions on necessary training schemes in the water sector due to, for example, the envisaged introduction of specific technologies or approaches.

5.4.2 Obstacles at National Level

The obstacles that are grounded within the superordinate system already indicate the general difficulties that are related to a problem- or demand-orientated training system and thus, to enhancements in the water sector. The superordinate system sets the boundaries for the wide influence of the national level (see Fig. 5.4) on the regional or company levels and decisions.

5.4.2.1 Training Demand Assessment Does Not Reflect Specific Demands

Although the government changed the former top-down demand assessment procedure to a bottom-up approach, it is still not regarded as demand-responsive. The training demand is mainly defined based on the job descriptions that were prepared

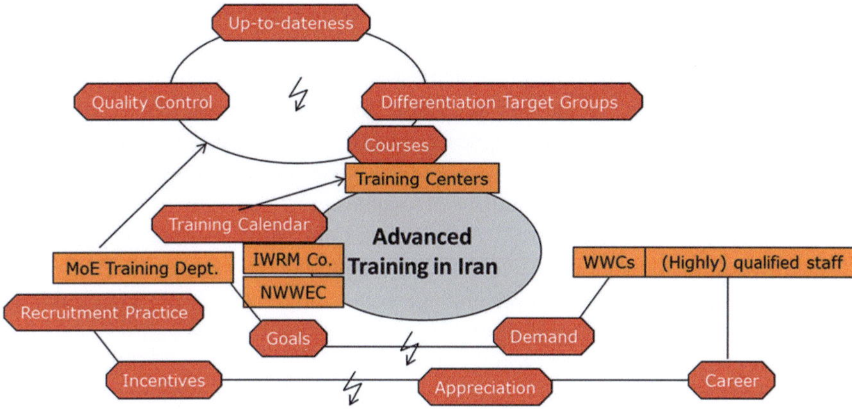

Fig. 5.4 Obstacles at the national level

during the training sector reform. Usually, human resources and the training bureau, together with the managers, agree upon the training demand of employees. Their recommendation, however, needs to be approved by the superior organization (NWWEC or IWRM Co.). The company staff at operational level is not involved in the demand assessment, but can only chose courses offered in the annual training calendar.

It was noted by interviewees that this system of training calendar development does not take into account the different local conditions and problems that need specific solutions and therefore the provision of particular training. Only very general courses are adopted in the national training calendar. Very specific training requests would often not be incorporated into the national calendar, and are therefore not offered by governmental training centres.

It is, however, possible for companies to propose specific courses to be subsequently incorporated into the national training calendar. But the procedure of approval by the Ministry of Energy takes several months and is said to be cumbersome.

5.4.2.2 Coding System Nurtures Lack of Motivation and Causes Loss in Quality

The coding system seems to play a central role with regards to the motivation of employees in attending training courses, but also influences the quality of courses. Usually, only governmental training centres offer courses that are codified by the Ministry of Energy, and only credit points or certificates of these courses can be collected for job promotions, rises in salary or higher educational degrees. It was stated by interviewees that these rewards are the major reasons for employees to attend the courses, which goes hand in hand with the statement that there is no intrinsic motivation based on individual or company-specific training demand.

Employees would often just attend a training course to collect credit points or certificates.

It was also said that an attempt to increase motivation by introducing the above mentioned rewards led to another problem: As these rewards require a certain amount of training hours, some courses were just extended, e.g. from 6 to 18 h in order to fulfil the requirements but without adding substantial training content. This again has led to the decreased quality of courses, and explains why participants perceive training as "wasted time". At the same time, these courses become more expensive for the clients (WWCs or RWCs) but provide less benefit which in turn leads to declining motivation for the management to appreciate or promote training.

Companies sometimes organize off-the-cuff training directly with training providers rather than following the formal procedure via the Ministry of Energy which is said to be inflexible and time-consuming. This way, though, participants are not able to collect credit points for their attendance which, again, dilutes their motivation.

5.4.3 *Obstacles at Operational or Company Level*

5.4.3.1 Lack of Authority Leads to Lack of Proactivity

Interviewees stated that the overall centralized organization and structure of the water sector in Iran causes a situation where subordinated companies "get used to receiving instructions" from their superior authorities and have neither the chance nor the motivation to be proactive (Fig. 5.5).

WWCs and RWCs have no legal basis for the development of financial or investment planning. In general, WWCs do not operate cost-efficiently and therefore their financial means for advanced training are limited anyway. The lack of authority to determine technological development or to introduce new technology is regarded as another reason for the limited motivation of management to invest in advanced training for their employees.

Some interviewees stated that problems in the training sector also arise as a result of established recruitment processes that are still widespread in Iran. The Ministry of Environment usually decides on the employment of people for the subordinated companies. Often, these people are not sufficiently educated in the respective field but job-specific training has to be facilitated and paid by the companies, which is often too expensive. The individual qualification of the selected employees is not evaluated and therefore not matched with the company's specific demand.

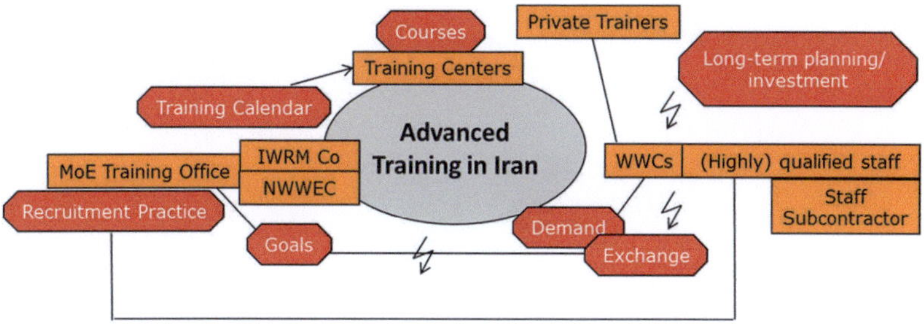

Fig. 5.5 Obstacles at operational level

5.4.3.2 Human Resources Management Does Not Promote Advanced Training

As has been described, advanced training is not necessarily appreciated by employees or management of water sector organizations. Interviews suggest that this is also reflected in the attitude of HR and training departments which are main actors in the field of training as they contribute to the training demand assessment. Interviewees said that HR managers themselves lack know-how of the latest technological trends, but especially of effective HR development practices. Adequate knowledge of demand assessment approaches or motivation techniques would be required to fulfil the actual task of bridging company and staff demands.

The theoretical approach of training evaluation in Iran is the Kirkpatrick Model[6] which does not only evaluate the training course itself but also its impact on a company's operation. To assess these impacts would be an original task for HR departments in order to facilitate a feedback loop, but it is currently not carried out.

Eventually, a lack of proper evaluation combined with a lack of knowledge about the employees' know-how and capabilities (as they are selected by the Ministry of Energy) and insufficient appreciation of training in general leads to deadlock. Hence, HR managers miss their opportunity of being the pivotal point of human resources development in their companies.

5.4.3.3 No Exchange of Experiences

A last point that was mentioned is the lack of an exchange of experiences between the individual companies. A know-how exchange seems to be lacking at the management level but also at operational level or between HR departments. Without any exchange about new technologies or (management) approaches there is also little benchmarking which would probably engage people to invest more in human resources.

---

[6]http://www.kirkpatrickpartners.com/OurPhilosophy/TheKirkpatrickModel (accessed May 2016).

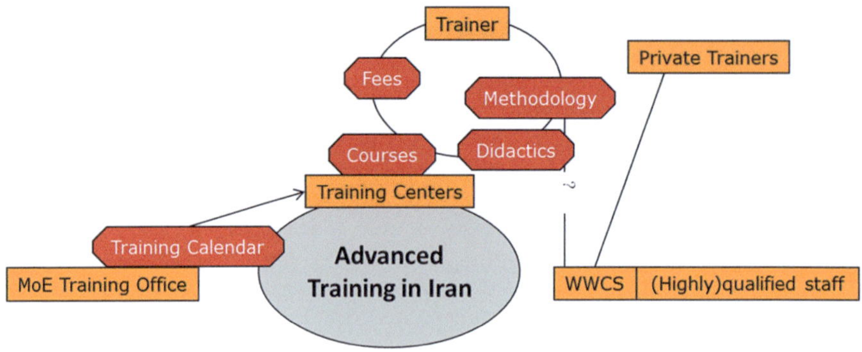

Fig. 5.6 Obstacles at training centre level

5.4.4 Obstacles at Training Centre Level

There are various obstacles for sustainable development of the advanced training structure in Iran's water sector which are grounded or reflected in the training centres' organization. The following part only deals with obstacles of governmental training centres which are subordinate to the Ministry of Energy (Fig. 5.6).

5.4.4.1 Lack of Competition Leads to Decreasing Training Quality

A central argument of the interviewees for insufficient quality of training at governmental training centres was the lack of competition between training providers. As the training centres are subordinate to the Ministry of Energy, and as the ministry has great influence on the companies' choice of the training facility (around 90% of training of WWC and RWC staff is done at governmental training providers), there is basically no free market in the training field. This lack of independence and competition would inevitably lead to the decreasing quality of training courses. Privatization of the training centres, however, is said to be obstructed by the Minister of Energy.

5.4.4.2 Providing Fixed Courses Causes Inefficient Binding of Resources

Governmental training centres have to provide all courses that are set in the national training calendar, around 1200 per year. This means that not only teaching material but also facilities and trainers that are trained accordingly, must also be provided throughout the entire year. This approach produces costs for the training centres even for courses that would, under conditions of market competition, probably be no longer offered.

5.4.4.3 Lack of Adequate Allocation of Financial Resources

There were different statements with regards to the allocation of financial means for training. Training centres condemned their minimal financial autonomy with regards to cost calculation and a weak payment moral of the water companies which would lead to additional financial problems. Fees for the training courses are also determined by the Ministry of Energy. Inadequate financial means, again, would lead to poor equipment and eventually inferior training quality.

Water companies, on the other hand, stated that training courses at governmental training facilities are more expensive than at independent providers. They criticized the poor allocation of finances, for example for large training rooms, rather than for well-educated trainers or demonstration facilities for hands-on training.

5.4.4.4 Poor Training with Regards to Didactics and Methods

Trainers have difficulties providing training with practical elements (hands-on seminars) and interactive methods. This may be due to the fact that trainers are often not practitioners themselves, or due to the lack of competition which does not motivate for different or exceptional training methods. But it may also be a problem of deficient equipment for more practical training. Often, courses are held in a one-dimensional manner, with the participants being a passive audience. Regardless of the reason, lecture-dominated training is not capable of enabling participants for autonomous problem definition and solution in their day-to-day business. In general, trainers that are not employed by the training centres change too often to allow for adequate quality control.

5.5 Conclusion and Recommendations

During the interviews and workshops, initial conclusions and ideas for overcoming the obstacles were discussed. It can be stated that the demand assessment is still regarded as being too top-down and supply-side oriented while there is little consideration of the specific demands at operational level. Motivation of employees to participate in training is strongly connected to the prospect of collecting credit points for rewards, and there is little intrinsic motivation like individual efforts to improve company business and operation.

A lack of competition between the training centres leads to declining quality of courses with little practical input and outcome. Evaluation of training is carried out by the training centres as a loose end process. There is no feedback loop back into the training centres and therefore there are no resulting measures for improving the courses. The image of training, thus, can only be improved if people have confidence in good training quality with benefits for their companies (Fig. 5.7).

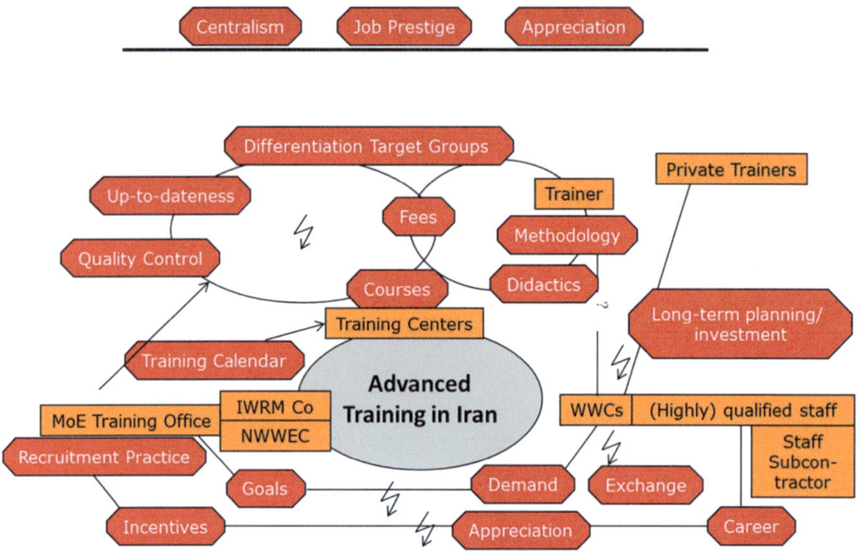

Fig. 5.7 Overview of all obstacles for the development of sustainable advanced training

Recommendations developed in close cooperation with the Iranian dialogue partners can be concluded as follows:

- Promote the privatization or at least greater independence of governmental training centres to increase competition, and increase the importance of private training centres compared to governmental training centres;
- Restructure and re-engineer governmental training centres with regards to management, efficiency, complexity, processes and methods;
- Increase the quality of training courses, add new courses for new and innovative technologies and strategies in the water sector;
- Improve training effectiveness evaluation by applying at least three of the four steps of the Kirkpatrick model: "reaction" (assessment of the participants' satisfaction), "learning" (knowledge demonstration or test), "behaviour" (evaluation of knowledge transfer 'on-the-job') and "result" (evaluation of the company's performance);
- Increase individual training motivation by linking training to overall company success;
- Update training courses for HR departments in the field of new human resources development methods;
- Introduce mentoring and coaching of managers in the field of human resources management;
- Develop a company-specific demand analysis for company staff and sub-contractors;
- Continuously evaluate individual qualification and competencies by employer;

- Introduce measures for exchange of experiences at operational and management level, like meetings and workshops and participatory strategic development processes;
- Introduce processes to include employees' personal demands in demand assessment procedures (e.g. regular meetings of employers and employees).

These recommendations shall serve as starting points for identifying strategic approaches for overcoming the described obstacles. The definition of strategic approaches which will be applied to a German Iranian Competence Center for Water and Wastewater Management (GICC) that is in the planning process will be part of the second project phase.

References

Bruns E, Köppel J, Ohlhorst D, Schön S (2007) Konstellationsanalysen als Brückenkonzept für die Innovationsforschung. In: Hof H, Wengenroth U (eds) Innovationsforschung: Ansätze, Methoden, Grenzen und Perspektiven. Lit Verl, Münster, pp 143–159

Ibisch R, Leidel M, Niemann S, Hornidge AK, Goedert R (2016) Capacity development as a key factor for integrated water resources management (IWRM): lessons learned from a series of applied research projects. In: Borchardt D, Bogardi J, Ibisch R (eds) Integrated water resources management: concept, research and implementation, pp. 335–373. Springer, Cham. doi:10.1007/978-3-319-25071-7

Mohajeri S, Nuñez von Voigt T (2011) Integrated water resources management in Iran. Isfahan, Iran: successfully shaping international technology and knowledge transfer. In: Water wastewater technology wwt international. Trade journal for water and wastewater management. Special issue, Berlin: HUSS-MEDIEN GmbH, pp. 29–32

Schön S, Mohajeri S (2008) Identification and analysis of key issues which constrain progress in the Iranian wastewater sector, Report for the World Bank

Schön S, Kruse S, Meister M, Nölting B, Ohlhorst D (2007): Handbuch Konstellationsanalyse. Ein interdisziplinäres Brückenkonzept für die Nachhaltigkeits-, Technik- und Innovationsforschung (Handbook of constellation analysis. An interdisciplinary bridging concept for sustainability, technology and innovation studies). Munich

Schön S, Nölting B, Meister M (2004) Konstellationsanalyse. Ein interdisziplinäres Brückenkonzept für die Technik-, Nachhaltigkeits- und Innovationsforschung (Constellation analysis. An interdisciplinary bridging concept for sustainability, technology and innovation studies). Discussion Paper 12/2004, Technical University Berlin

Part III
Current and Future Water Availability and Use

Chapter 6
Current and Future Agricultural Water Use in the Zayandeh Rud Catchment

Wolf Raber

6.1 Introduction

One of the major achievements in the history of civilization is the development of agriculture. Archaeological research shows the origin of agriculture to be in southwest Asia, the so called Fertile Crescent or Cradle of Civilization. The western part of Iran along with the Zagros Mountains is part of this area with evidence for agriculture dating back 12,000 – 9800 years (Riehl et al. 2013).

Agriculture is still a highly important sector in modern Iran and its economy. The significance of agriculture is emphasized by the fact that in 2010 about 20% of all employed people were working in agriculture (Statistic Center of Iran, 2015). Especially in rural areas, 94% of the farms are family owned businesses with arable land of 0.1 ha- 10 ha (Nikouei and Ward 2013). Export goods are also produced, the three major agricultural export products in 2010 were pistachios, apples and raisins, but also high value spices are exported, while maize, meat and raw sugar were the commodities with the highest amount imported to Iran (FAOSTAT, 2010). In the 1990s agriculture was the fastest growing economic sector in Iran with a quarter of the national GDP, this decreased in the following years due to severe drought to a GDP share of only 14% in 2005 (Stads et al. 2008).

Its importance is manifested in the role of agriculture within the 20 year Perspective Plan of the Iranian Government.[1] Important aims of the Perspective Plan are:

- Implementing water balance projects along with improving irrigation methods;
- Increasing agricultural water productivity;

[1] http://www.cila.ir/ (Accessed: May 2016).

W. Raber (✉)
inter 3 GmbH Institute for Resource Management, Berlin, Germany
e-mail: raber@inter3.de

S. Mohajeri, L. Horlemann (eds.), *Reviving the Dying Giant*,
DOI 10.1007/978-3-319-54922-4_6

- Developing new agricultural land;
- Protecting and improving agricultural productivity;
- Achieving self-sufficiency.

This contribution deals in particular with agriculture in the Zayandeh Rud catchment within Isfahan province in central Iran. The region is strongly influenced by two aspects: the city of Isfahan and its urban and industrial influence and the rural side with a strong emphasis on agriculture. Agriculture in the catchment strongly depends on water supply for irrigation. The available water is distributed between urban, industrial and agricultural sectors and the environment. Agriculture consumes by far the largest proportion of the surface- and groundwater resources in the Zayandeh Rud basin with up to 90%. Cultivation of high water demanding crops and low irrigation efficiency of 34–42% contribute to the high agricultural water demands (Murray-Rust et al. 2002).

The government of the province of Isfahan developed policies for the agricultural sector in accordance with the 20 year Perspective Plan. The main development policies include an increase in application of modern cultivation methods in agriculture and greenhouse crops. Equally important is the introduction and integration of cropping patterns combined with agricultural management along with the creation of new product organizations, insurances and specific investment models. One very important point in the development plan is the implementation of methods for combating desertification and preventing the spread of the bordering deserts.[2]

A high impact problem in the Zayandeh Rud catchment is water scarcity. Since the year 2000, the agricultural sector in the Zayandeh Rud basin has been suffering from drought again and again. Depending on water availability, characteristics of agricultural cultivation have been changing from year to year.

One major goal of the Iranian-German IWRM Zayandeh Rud research project is to develop a Water Management Tool (WMT) (see Chap. 12). In regard to agriculture, the WMT is supposed to model and forecast the spatial distribution of future agricultural water demand under different development scenarios, as well as to model the impact of different scenarios of water availability on the agricultural sectors. For calculating the temporal and spatially differentiated water demand of agriculture in the basin the internationally accepted approach suggested by the Food and Agricultural Organization (FAO) in the Irrigation and Drainage Paper No. 56 (Allen et al. 1998) is applied in the WMT. Next to a range of climate data (see Chap. 13) for a reliable model, a solid data base with the following information is necessary:

- Location and extent of cultivated and irrigated areas;
- Specification of types of cultivated crops and orchards in these areas;
- The sources of irrigation water and applied irrigation methods;
- A crop calendar with sowing date and growing periods of different crops;

[2] http://www.agri-es.ir/Default.aspx?tabid=5865 (Accessed: May 2016).

- Crop specific data for computing the crop water requirements according to FAO Paper No. 56, like crop coefficient, crop development phases, root depth, maximum vegetation height and depletion fraction.[3]

In this contribution the focus lies on presenting insights on these aspects for the normal wet year 2006 as a reference year. By choosing an average wet year as the reference year, it can be assumed that a maximum of cultivated area is under irrigation. Therefore, this year may be used as a demand base for other years with possible deficits in supply.

Based on the data presented in this contribution and with the modeling options of the WMT, an ambitious strategy development of the agricultural sectors can be conducted in the IWRM Zayandeh Rud research project. The overall goal is the development of adapted and effective measures for agricultural transformation in the catchment. The ongoing work to this end is presented in Sect. 6.7 of the document at hand.

In general, it has to be noted that the figures presented are the product of an intense data collection and analysis process. There are several studies available on agricultural activities in the Zayandeh Rud basin (e.g. Zayandab Consulting Engineers Co 2008, Molle et al. 2004 and Sally et al. 2001) which have a focus on specific regions or aspects of agriculture. The greatest challenge was to create one harmonized database with a common temporal and spatial format, integrating relevant information from different institutions and administrative levels. In this regard, it has been proven as particularly difficult to format data for different irrigation networks as spatial reference, since Iranian raw data are available either with an administrative reference (national, province, county or village scale) or hydrological reference (basin or hydrological subunit) which are not directly compatible with each other.

6.2 Location of Irrigated Areas

The Zayandeh Rud Catchment is part of the Gavkhuni basin. The natural appearance of the area is dominated by the ("life giving") Zayandeh Rud river, which originates in the Zagros Mountains, where snow and rain fall are regular with an average of 1700 mm (Molle et al. 2009), crosses the most populated and industrialized areas around Isfahan and ends after flowing 400 kilometers into the Gavkhuni wetland surrounded by desert which receives only around 100 mm rain per year. According to Faramarzi (2014), the Zayandeh Rud catchment can be divided into three main climate zones:

- Semi-humid in the west uplands;

[3]Due to the extent and complexity of the last mentioned, very theoretical sets of crop specific information for computing crop water requirements, these are not part of the text at hand.

- Semi-arid in the central;
- Arid climate in the east lowlands.

The minor amount of rainfall, particularly in the fertile and intensively cultivated central and eastern part of the catchment, makes irrigation a basic requirement for agriculture. Surface water from the Zayandeh Rud represents the main source for irrigation water in the catchment. The Zayandeh Rud is controlled by the Chadegan Dam, which retains a water reservoir of 1500 million m^3 water (Hoogesteger 2005). Other water sources used for irrigation are percolated surface water from the river exploited by bank wells, groundwater from wells and qanats as well as spring water. Rain-fed agriculture plays a role only in the western part of the catchment and is disregarded in this document due to its overall insignificance. The assumption is taken that all presented cultivated areas are being irrigated.

With the importance of surface water supply for agricultural activities, it was found that almost all cultivations are located within the command area of different irrigation networks. Furthermore, conjunctive use of surface and groundwater is common for farmers and shallow ground water resources in the areas of irrigation networks stand in a direct relationship with surface water supply to the networks (see Chap. 11 Roodasht). Therefore, the areas of irrigation networks have been chosen as geographic reference for the water management model, which each include one single balance point of water withdrawal at the headwork of each irrigation network.

By applying and combining geographical data in form of shape files for geographical information systems (GIS) provided by the Isfahan Water Company and aerial pictures with vegetated surfaces in the catchment, 15 irrigation networks could be identified and mapped.[4] The identified irrigation networks are spread throughout the whole catchment mainly along the Zayandeh Rud River and comprise modern as well as traditional channels.

6.3 Extent and Types of Cultivated Crops

In this chapter the extent and type of cultivated areas in the identified irrigation networks is presented. The data on cultivated areas is derived from datasets on cropping patterns compiled by the Agricultural Organization Isfahan.

In the wet reference year the total cultivated area in the Zayandeh Rud catchment summed up to 225 T ha. In close collaboration with the Agricultural Organization Isfahan the cultivated areas could be distributed to the boundaries of identified irrigation networks. In Fig. 6.1 the locations of identified irrigation networks

[4]These are Kordolia (morghab spring), Karevan up, Karevan down, Lenjanat Down (anhar sonati downstream gauge Lenj), Lenjanat Up (anhar sonati upstream gauge Lenj), Mahyar Total, Nekouabad right, Nekouabad left, Borkhar, Abshar right, Abshar left, Rudasht north, Rudasht south, Faridan and Fereydunshahr, Chadegan.

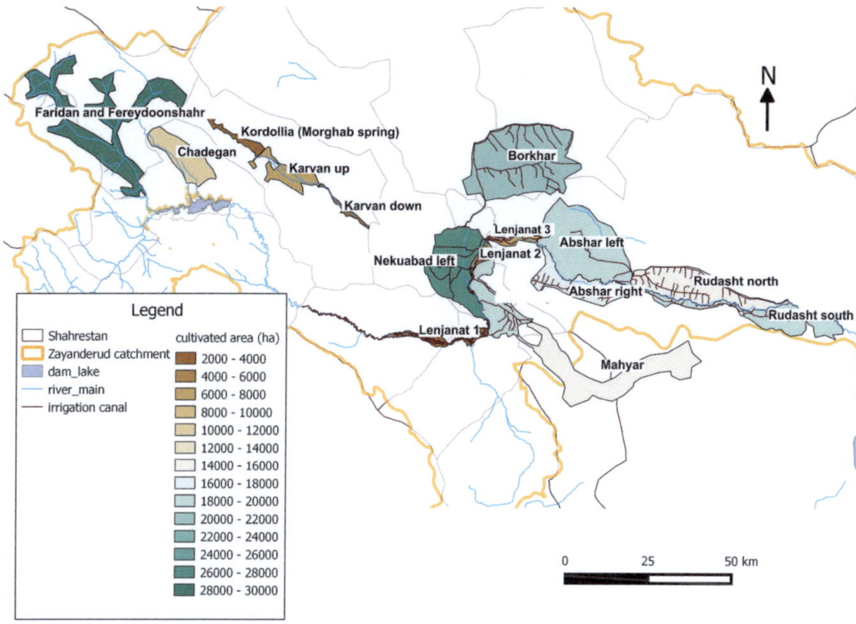

Fig. 6.1 Cultivated area in the irrigation networks in Zayandeh Rud Catchment (reference year: 2006)

including the specific extent of cultivated areas is presented on a map with the boundaries of the Zayandeh Rud catchment, main river stream and administrative units (Sharestan). Isfahan city is located in the central area between Nekouabad left, Borkhar, Abshar left, around Lenjanat 3.

It stands out that irrigation networks around the Khamiran dam (Kordolia, Karevan) (west of Isfahan) as well as Lenjanat are comparably small and have very little production. The other identified irrigation networks comprise 90% of cultivated area in the catchment. The two largest irrigation networks Nekouabad close to Isfahan city and Faridan and Fereydunshahr in the western highlands above the dam comprise 58 T ha, 26% of the basin wide cultivated area. Despite the large extent of Faridan and Fereydunshahr, cultivated areas in the upstream semi-humid western uplands account for only 24% of catchment wide cultivation.

With further analysis, the cultivated areas could be separated into areas cultivated with crops, orchards and modern greenhouses. Cultivation in greenhouses is done mainly for high value vegetables, but in 2006 covered less than 150 ha and is therefore excluded from further analysis due to its insignificance. Table 6.1 states the cultivated areas in the identified irrigation networks,[5] split into crops and

[5]In the table, the irrigation network "Dehaghan" is presented which is located somewhere in the south of the Zayandeh Rud basin. This network could not be located exactly and is therefore excluded from the maps presented.

Table 6.1 Cultivated area [ha] in different irrigation networks distributed in crops and orchards in the year 2006 (source: Agricultural Organization Isfahan)

	Total cultivated area		
Irrigation network	Crop [ha]	Orchard [ha]	Sum [ha]
Kordolia (morghab spring)	1943	976	2919
Karevan up	4737	2935	7672
Karevan down	1275	1195	2470
Lenjanat down	4690	1944	6633
Lenjanat up	985	1046	2031
Mahyar total	14,184	152	14,336
Nekouabad right	17,119	1626	18,745
Nekouabad left	22,335	5499	27,834
Borkhar	20,448	1092	21,540
Abshar right	15,878	884	16,762
Abshar left	18,678	357	19,035
Rudasht North	15,445	50	15,495
Rudasht South	18,340	60	18,400
Faridan and Fereydunshahr	27,790	2160	29,950
Chadegan	10,356	1142	11,498
Dehaghan	8558	1071	9629
Total	202,761	22,187	224,949

orchards. The presented figures indicate that 90% of the cultivated area is covered by crops and only 10% by orchards, where orchards are mainly located in the western and central part of the catchment.

On the 203 T ha cultivated with crops in the year 2006, different varieties are found (see Fig. 6.2). Wheat is the prevailing crop and is cultivated on 70 T ha, which represents 31% of the total cultivated area in the catchment. When looking at the agricultural area cultivated with the six most predominant crops, namely wheat, barley, alfalfa & sainfoin, rice, potato and maize, the area sums up to 162 T ha which represents 72% of the catchment wide cultivated area. Other crops like vegetables, legumes and onions are cultivated on land between 4 and 5 T ha, clover, kitchen garden, sunflowers and sugar beets cover areas between 2 and 4 T ha, whereas the rest is cultivated on areas smaller than 2 T ha.

Besides the crops mentioned above, a number of perennial trees and bushes are cultivated in orchards. Regarding the distribution of plants in orchards depicted in Fig. 6.3, one can see that almonds and grapes are planted on the largest areas with each above 4000 ha. In sum, stone fruits, grain fruits and apples are cultivated on an area larger than 7300 ha and are merged with the category fruits for analysis later due to its proximity of cultivation characteristics. Also walnuts and olives, where most of the other products with wild olives are accounted for, are popular amongst farmers whereas high value export goods, namely pistachios and saffron, are cultivated on 333 ha and 109 ha only.

In the overall analysis of the distribution of cultivated areas in the irrigation networks, it can be found that cultivated crops are not distributed evenly over the

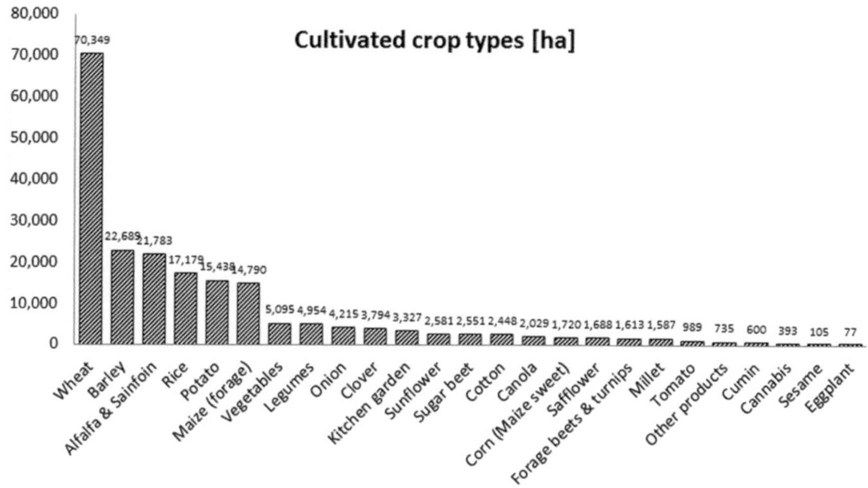

Fig. 6.2 Identified crops on farmland in the Zayandeh Rud Catchment in 2006 (source: Agricultural Organization Isfahan)

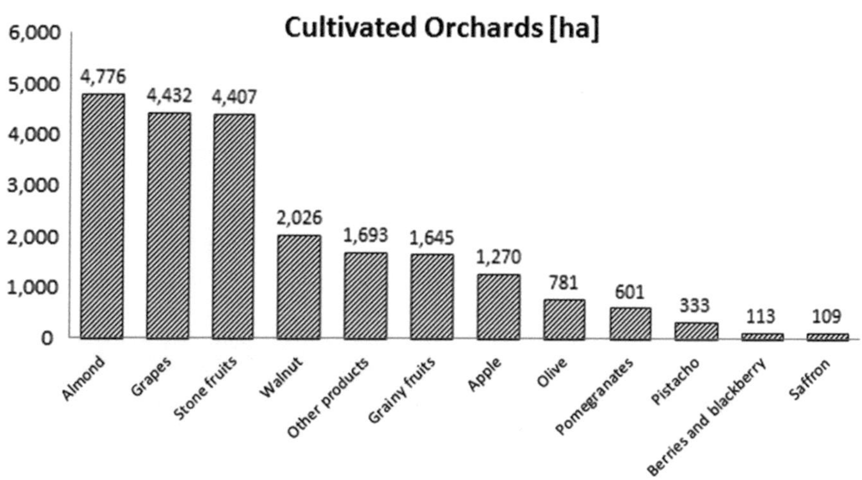

Fig. 6.3 Identified orchards in the Zayandeh Rud catchment in 2006 (source: Agricultural Organization Isfahan)

catchment area, but some irrigation networks have a prioritized production for specific crops and orchards. For example the following dominant productions stand out:

- Wheat in Rudasht north and south whereas also the main part of canola is in Rudasht south;
- Alfalfa & sainfoin and potato in Faridan and Fereydunshahr;
- Rice and fruits in Nekouabad left;
- Almonds and grapes in Karvan up.

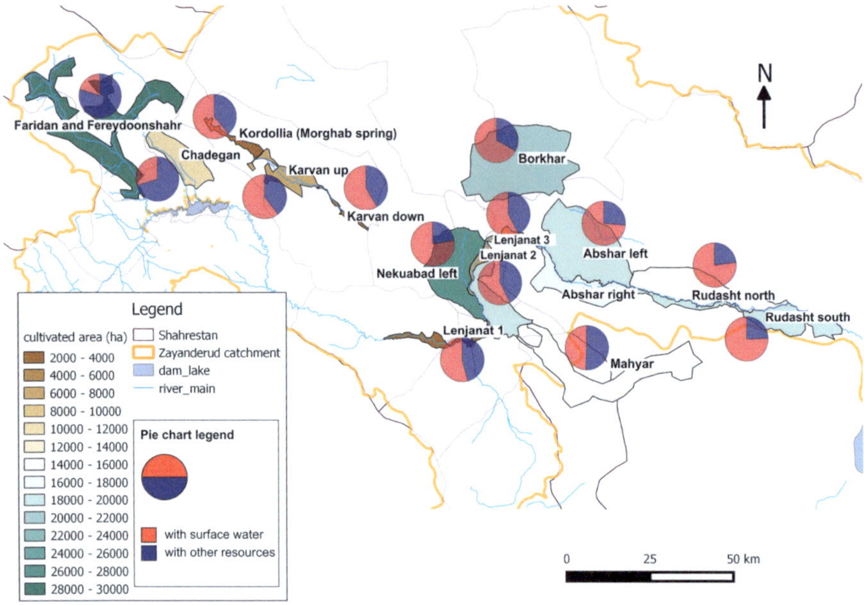

Fig. 6.4 Distinction of irrigation water sources in irrigation networks of Zayandeh Rud catchment

6.4 Irrigation Water Sources

The water sources for irrigation are either surface water of the Zayandeh Rud or other sources like groundwater from deep or shallow wells, bank wells along the river or springs and qanats.[6] By analyzing data from the Agricultural Organization Isfahan and the Regional Water Board Company, sources for the applied irrigation water could be assigned to each irrigation network for the reference year 2006.

Data show that approximately 60% of the total cultivated area is irrigated by surface water and 40% by other resources. Regarding water sources, no difference could be found for crops and orchards, but for the geographical location. In Fig. 6.4 a map with the distinction of irrigation water sources in the different networks is presented for the Zayandeh Rud Catchment. The irrigation networks are colored according to their extent of cultivated area and the pie chart shows the proportion of irrigation water coming from surface water and other water resources. It stands out that usage of other water resources is happening mainly in the western, upstream region of the catchment (mainly Faridan & Fereydunshahr and Chadegan). In the western, mountainous part of the basin, deep wells and qanats as well as some springs are dominant water sources. In irrigation networks along the river (Lenjanat and Abshar right) bank wells are found widely, whereas in the centrally located irrigation networks (Nekouabad, Borkahr and Abshar) deep wells are popular. In

[6]Traditional infrastructure for taping groundwater.

the downstream area (Roodasht) as well as in Lenjanat also a lot of shallow wells are located.

6.5 Irrigation Methods

In the Zayandeh Rud Catchment the main fraction of cultivated area is being irrigated traditionally by flood and furrow irrigation. However, in the past year modern irrigation methods have been introduced on some plots. In this paragraph data on irrigation methods in the catchment will be presented, which distinguish between two categories of irrigation methods: (a) pressurized (e.g. sprinkler and drip) and (b) gravity (e.g. flood and furrow).

Figure 6.5 shows that mainly field crops like potatoes, wheat and alfalfa as well as almonds in orchards are irrigated with pressurized methods. Analysis of different types of pressurized irrigation shows that 100% of sprinkler irrigation is applied on crops and 85% of the drip irrigation is applied on orchards. Remaining drip irrigation is used with 7% for green spaces and an additional 7% for crops (mainly kitchen garden and potato).

In 2006 about 10% of the cultivated areas in the whole basin have been irrigated by pressurized irrigation. In some areas modern irrigation methods find more application than in others (see Fig. 6.6). In Fereydonshahr and Fereydan upstream of the dam, almost 30% of the cultivated areas are irrigated by pressurized irrigation and also Chadegan shows an above average percentage application (18%) of pressurized irrigation methods. In regard to the previous chapter, it stands out that in the irrigation networks with high uptake of pressurized irrigation methods a large share of irrigation water comes from deep wells and quants. This correlation is consistent with the fact that pressurized irrigation methods require a constant flow of high quality water.

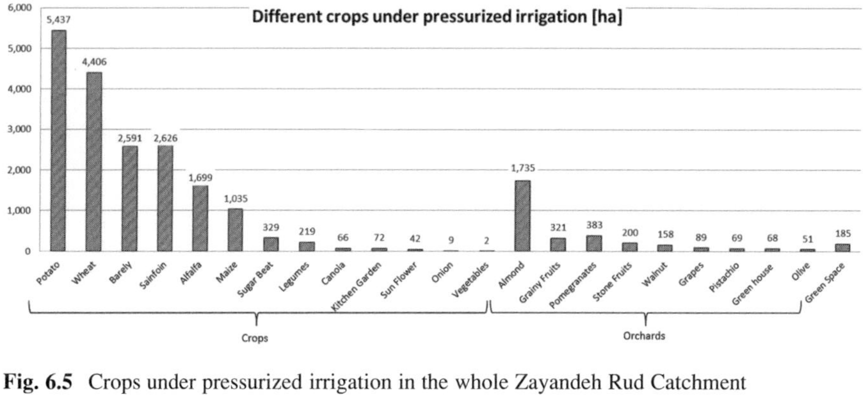

Fig. 6.5 Crops under pressurized irrigation in the whole Zayandeh Rud Catchment

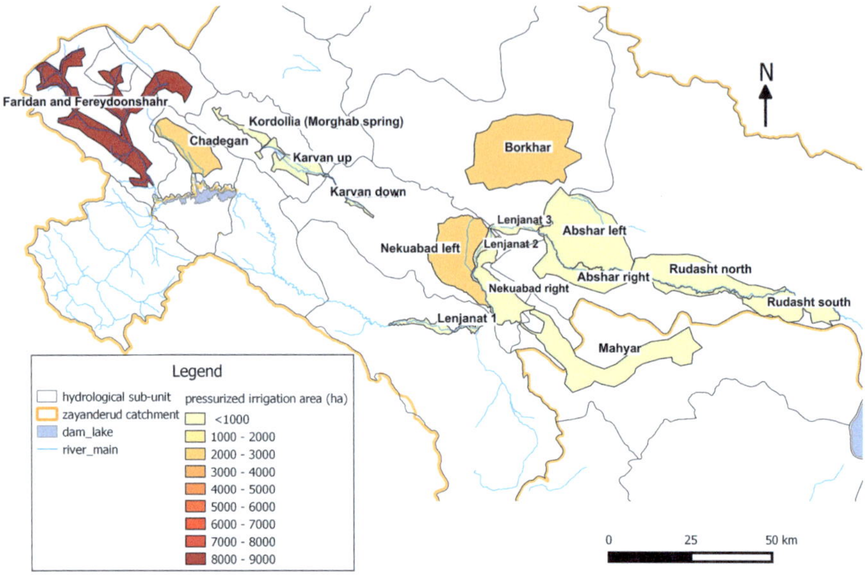

Fig. 6.6 Areas irrigated with pressurized irrigation methods distributed to identified irrigation networks (2006)

6.6 Crop Calendar

To calculate the time-related irrigation water demand in the Zayandeh Rud catchment, a crop calendar with information on planting dates and cultivation periods for all relevant crops and orchards has been created. The calendar is based on extensive data on cultivation periods for each irrigation network by the Agricultural Organization Isfahan. Since the irrigation networks are located in different climate zones (semi humid western uplands, semi-arid central area and arid eastern lowlands) cultivation length as well as planting and harvest dates of the crops and orchards vary throughout the whole basin. In a time-consuming process, the datasets could be simplified and clustered.[7] In Fig. 6.7 one united crop calendar for the whole basin is presented together with average climate data for Isfahan city. It has to be noted that the crop calendars used for the Water Management Tool are more diverse and location specific.

[7]Stepwise simplification has been performed by firstly setting all sowing dates on either the 1st or the 15th of a month. Second, the variation in the total growing period is reduced. For crops with large variations in total growing periods, subgroups with average growing periods have been created with a subsequent adaption of the harvest date. Furthermore simplifying assumptions on the data on orchards were carried out, as orchards are exclusively composed of perennial variations and trees, a vegetation period with irrigation demand was estimated for these for computing irrigation water demand.

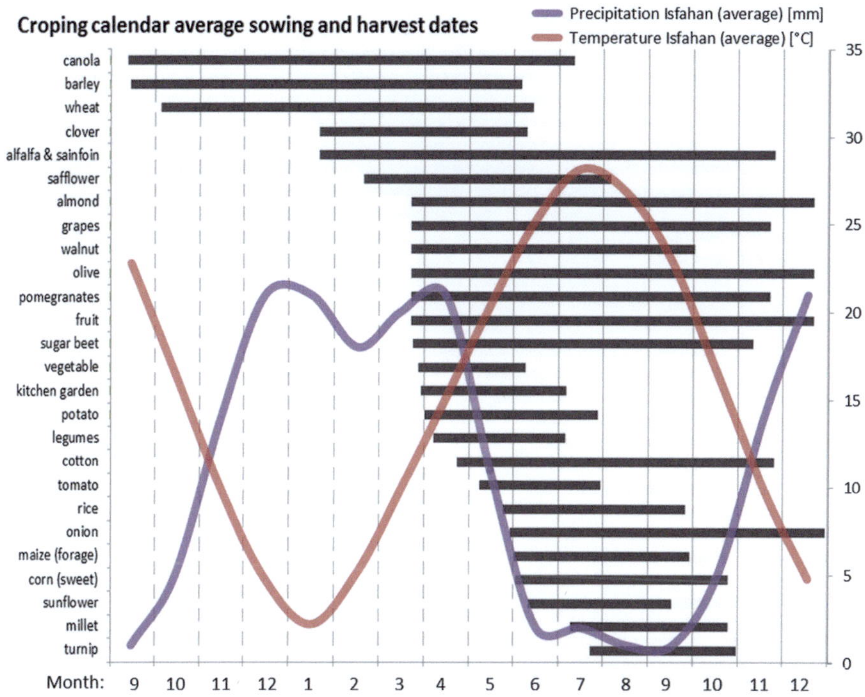

Fig. 6.7 Average crop calendar in the Zayandeh Rud catchment with average monthly precipitation and temperature for Isfahan city (Gregorian Calendar) (source: Cultivation and Harvest Calendar, climate-data.org)

The figure shows that the highest density of parallel cultivation occurs in spring around April and May (months 4 and 5) at the end of the wet season and amid rising temperatures. Besides canola, barley and wheat as well as clover, which have their main cultivation periods in winter times, with comparably high precipitation and low temperatures, all other crops have their main cultivation period in the summer when temperatures and evapotranspiration are high and almost no rainfall can meet the water demand of plants. This description highlights the agricultural dependency on irrigation water and the need for efficient, fair and sustainable water distribution and management on the one hand and approaches towards a transformation of the agricultural sector towards a modern and water efficient undertake.

Besides the timing, the length of cultivation periods determines the dependency of irrigation water and vulnerability towards droughts, overall water demand and requirements for maintenance by farmers. In Fig. 6.8, the large variability in average cultivation period of the cultivated varieties is presented. Considering the extent of the cultivated area of specific crops in Sect. 6.3, it stands out that the crops and orchards with the longest cultivation periods (>250 d) are also the most popular varieties in the catchment. Alfalfa and sainfoin, canola, fruits, olives, almonds, barley and wheat account for almost 60% of the total cultivated area within the Zayandeh Rud catchment.

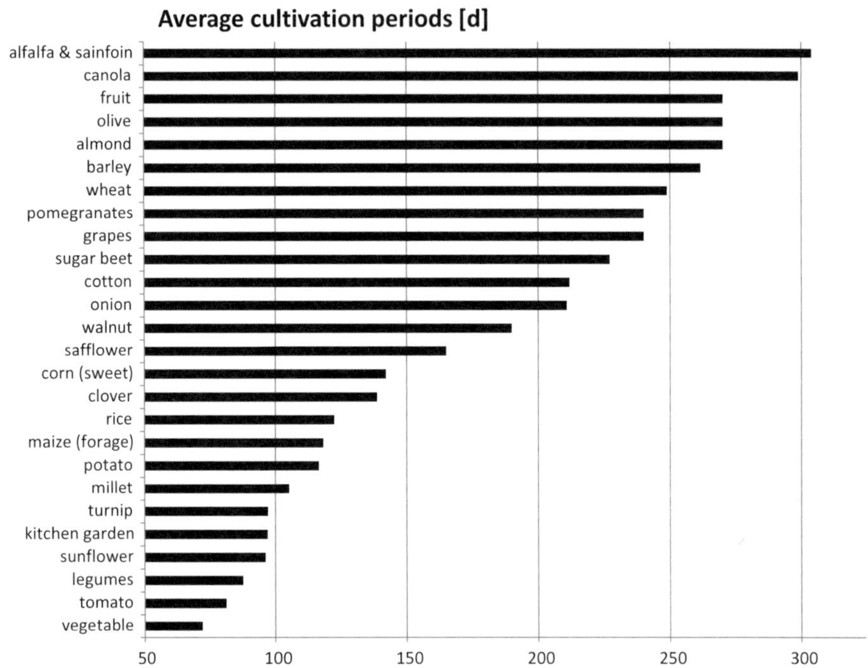

Fig. 6.8 Average cultivation periods (source: Cultivation and Harvest Calendar)

6.7 Agricultural Transition

The data presented show that the agricultural sector is one of the most complex and at the same time most important sectors for water management in this region. It is certain that the sector will be challenged by future water shortage due to climate variations and growing water demands of other sectors like industry, growing population and an increased demand of other farmers due to the development of arable land. The transfer of water to other provinces adds to local water shortage (Felmeden et al. 2014). Next to scarce surface water, also salinization of soil and water resources (Droogers et al., 2000), groundwater depletion and soil erosion are serious problems threatening agricultural production in the Zayandeh Rud basin.

Based on the presented findings and supported by the modeling options of the Water Management Tool, strategies for shifting the agricultural sector towards more efficient land and water exploitation for food and income production are currently being developed in the second phase of the Iranian-German IWRM Zayandeh Rud research project.

A strategy for agricultural transformation in the Zayandeh Rud catchment is being developed in three main fields of action:

- *Participatory development of recommendations ("Citizen's Juries" workshops).* This sub-project aims at developing strategies for conservational handling of water resources (in qualitative and quantitative terms) in three workshops with relevant stakeholders (farmers and decision-makers). The active involvement of local stakeholders is regarded as an essential measure for developing accepted and sustainable solutions that actually benefit the people affected. The workshops will be conducted using the innovative "Planning Cells" method (Dienel 2002). In a four-day workshop, two groups of 20-25 people will discuss different possible measures for agricultural transformation like measures to improve economic water productivity (greenhouses or alternative crop choice) or measures to improve crop water productivity (irrigation methods, mechanization or organizational innovations). Prior to the discussion of each possible measure, experts will inform them about the potential impacts of each measure. In small working groups and in plena participants will evaluate and prioritize the measures. The final product will be a report with statements and recommendations that will be handed over to decision makers.

- *Technological measures: Innovative irrigation technologies and soil conditioner*. In two pilot projects strategies for the reduction of agricultural water use without lowering the production output in the catchment will be tested. In the first pilot project, an innovative irrigation technology is being introduced experimentally and its water productivity will be compared with the traditional irrigation systems. The technique consists of three interlinked units that communicate with each other: an irrigation module, a control module and a monitoring module. The irrigation module can be connected to the wastewater treatment plant and is able to quantify the nutritional composition of wastewater. The monitoring module assesses the nutritional composition of the soil. Using the calculated values and incorporating information on plant species and growth period, the control module determines the required amount of nutrients. Based on this, the required admixture of clear water can be calculated, which guarantees optimal nutrition of the agricultural crop. In the second pilot project, the application of innovative soil conditioner will help to identify adequate substrates for improved water efficiency and soil quality. The use of additives for improving the soil's water storage capacity will help to increase water productivity.
- *Using computer models for strategy development*. Gathered data on the hydrological system, recommendations developed in the workshops and results of the technological measures will be merged to develop and evaluate concrete strategies for agricultural transformation. The main tool for this sub-project is an innovative GIS-based water productivity model for farms that will depict changes in water demand and present different possible combinations of measures for adapted land use management.

This mix of measures and methods described will be integrated into the process of the development of a strategy for sustainable agricultural transformation. The Iranian partners are actively involved in the process. Information on feasible management options, again, will be fed into the Water Management Tool as the main instrument for integrated water management in the Zayandeh Rud catchment.

References

Allen RG; Pereira LS, Raes D, Smith M (1998) Crop evapotranspiration – Guidelines for computing crop water requirements – FAO Irrigation and drainage paper 56, Food and Agriculture Organization of the United Nations (FAO)

Dienel PC (2002) Die Planungszelle. Der Bürger als Chance. Westdeutscher Verlag, Wiesbaden

Droogers P, Salemi HR, Mamanpoush A (2000) Exploring basin scale salinity problems using a simplified water accounting model: the example of Zayandeh Rud Basin, Iran. IAERI-IWMI, Iran

FAOSTAT (2010) http://www.fao.org/faostat/en. Accessed May 2016

Faramarzi M (2014) Hydrological Modeling of Zayandeh Rud River Basin, final report, IWRM Isfahan Project, Isfahan University of Technology

Felmeden J, Schramm E, Sattary E, Davoudi A (2014) Agriculture in the Zayandeh Rud Catchment. ISOE Mater Social Ecol 40. ISSN 1614-8193, http://www.isoe.de/uploads/media/msoe-40-isoe-2014_01.pdf

Hoogesteger JD (2005) "Making do with what we have", Understanding drought management strategies and their effects in the Zayandeh Rud Basin, Iran. MSc Thesis. Wageningen University

Molle F, Ghazi I, Murray-Rust H (2009) Buying Respite: Esfahan and the Zayandeh-Rud River Basin, Iran River Basin Trajectories: Societies, Environments and Development. IWMI, Colombo

Molle F, Mamanpoush A, Miranzadeh M (2004) Robbing Yadullah's water to irrigate Saeid's garden: hydrology and water rights in a village of Central Iran. IWMI research report 80. International Water Management Institute, Colombo, Sri Lanka. vi, 43

Murray-Rust H, Salemi HR, Droogers P (2002) Water resources development and water utilization in the Zayandeh Rud basin, Iran. IAERI-IWMI Res Rep. 2002; 13, Colombo

Nikouei A, Ward FA (2013) Pricing irrigation water for drought adaptation in Iran. J Hydrol 503:29–46

Riehl S, Zeidi M, Conard NJ (2013) Emergence of agriculture in the foothills of the Zagros mountains of Iran. Science 341(6141):65–67

Sally H, Murray-Rust H, Mamanpoush AR, Akbari M (2001): Water supply and demand in four major irrigation systems in the Zayandeh Rud Basin, Iran. IAERI-IWMI Research Reports 8, Colombo

Stads G-J, Roozitalab MH, Beintema NM, Aghajani M (2008) Agricultural research in Iran: policy, investments, and institutional profile. ASTI-Country Report http://www.asti.cgiar.org/pdf/IranCR.pdf

Statistic Center of Iran (2015) www.amar.org.ir/english/Statistics-by-Topic/Labor-force#2222532-time-series. Accessed April 2016

Zayandab Consulting Engineers Co. (2008) Study of water resources and water consumption in the Zayandeh Rud Basin. Agriculture. Vol. 5, Isfahan

Chapter 7
Current and Future Industrial Water Use in the Zayandeh Rud Catchment

Wolf Raber and Shahrooz Mohajeri

7.1 Introduction

The Zayandeh Rud catchment is one of the most important industrial regions in Iran with a value added of approx. 14% of the GDP. Of strategic importance for Iran are large industry branches like steelworks, cement mills and oil refineries in the catchment. About 110 local mines supply industry with raw materials. Of the three main water using sectors – agriculture, domestic water services and industry – the latter consumes a proportion of 6–7% of the surface and groundwater resources of the catchment.

Industry in the catchment can be grouped into about 30 large industry units and about 13,000 small and medium sized commercial enterprises and manufacturers. Large industry is found exclusively in Isfahan province whereas smaller enterprises can also be found in the neighbouring province Charmahal va Bakhtiari. These approx. 180 small industrial units located outside of Isfahan province, while supplied by the Zayandeh Rud catchment, are of minor importance for the water management in the catchment.

In Isfahan province, around 315,000 employees work in the industrial sector. Small commerce and industry employ approximately 210,000 people and large industry 100,000, whereas about 90% are employed by two steel works alone. About 70% of the employees of small industries work in metal, non-metal and textile industries. An additional 3600 people work in mines and 1200 in industrial agriculture (e.g. large livestock and poultry farms) which is, according to Iranian legislation, a part of the industrial sector. From a (quantitative) water management perspective, mines play a minor role and are not discussed here.

W. Raber (✉) • S. Mohajeri
inter 3 GmbH Institute for Resource Management, Berlin, Germany
e-mail: raber@inter3.de

S. Mohajeri, L. Horlemann (eds.), *Reviving the Dying Giant*,
DOI 10.1007/978-3-319-54922-4_7

Although there are various surveys on the potentials and risks of further industrial development in the region (e.g. Bakhtiari et al. 2002; Reisi et al. 2011; Ghasemian et al. 2012), there is little known about the effects of – particularly future – industrial water consumption on the water resources in the catchment. Based on different criteria like groundwater tables, distance from water supplies, current land use etcetera Reisi et al. (2011) show that areas in and around Isfahan city suitable for industrial establishments are becoming less and less. Bakhtiari et al. (2002) found that resource efficiency in Isfahan province is very low posing a risk to further development. Yekom Consulting Engineers Company (2013a) suggest some measures for sustainable water use in industry, like reuse of industrial wastewater, based on data up to 2006. In three future scenarios Yekom Consulting Engineers Company (2013b) calculates the future water demand of industry based on the estimated increase of employees in the sector by 2041. Zayandab Consulting Engineers Company (2009) studied water resources and water use in industry in Zayandeh Rud catchment with the focus on industry in Isfahan province. The study indicates the main challenges related to water use in industry including the lack of statistical data and information as well unspecified amounts of water used in individual industries and hence pinpoints the basic reason for the survey presented in this chapter.

This chapter presents the findings of the industrial water consumption survey conducted within the IWRM Zayandeh Rud project. The survey's aim was to get an overview of current and future industrial water consumption and its relevance for IWRM implementation in the catchment, and to develop a coherent set of data which can be fed into the Water Management Tool (WMT, see Chap. 12).

7.2 Overview

The content of this chapter is selected and formatted according to available data and the specific requirements of the German-Iranian IWRM research project. It concentrates on the most relevant industries from an integrated water management perspective, which comprises more than 13,000 small, medium and large industrial units.

The presented figures originate from or are based on data selected in close collaboration with (1) local institutions like: Isfahan Water Board Company, Isfahan Water and Wastewater Company, Isfahan Industrial Organization and Industrial Settlement Organization, (2) interviews with representatives of large industrial units as well as (3) the review of reports on water consumption in the Zayandeh Rud catchment provided by Zayandab Consultancy Co. and Yekom Consultancy Co.

According to findings the total industrial water consumption in the catchment is approximately 200 MCM/a (reference year: 2006) including 48 MCM/a water consumption by industrial agriculture (reference year 2012). Next to a large quantity of small industries with minimal uptake of the overall water consumption, a few large water consumers, mainly from the metal, petrochemical and power generation sector, are intense water users. In consideration of the diversity of industrial units in the catchment, we divided the industrial water consumers into four groups:

- The 30 largest single industrial units with a consumption higher than 500,000 m^3/a per unit, which currently have a share of approximately 57% of the total industrial water consumption in the catchment;
- More than 10,000 small and medium sized industrial units are clustered into 29 large industrial settlements and zones. Each settlement is expected to have a consumption higher than 500,000 m^3/a by the year 2025. These industries currently account for approximately 14% of the industrial water consumption in the basin;
- About 3000 small scale industries within Isfahan municipal boundaries. They are currently supplied with approximately 10 MCM/a drinking water by the Isfahan Water and Wastewater Company, which amounts to approximately 5% of the industrial water consumption in the basin;
- Industrial agriculture with livestock and poultry farms, aquaculture and greenhouses which currently account for about 24% of the industrial water consumption in the basin.

Large single industrial units, industrial settlements and zones and industrial agriculture are presented separately in this document, since the respective data sources and analyses follow different concepts. The water consumption of small industries within Isfahan municipal boundaries are expected to be constant and only represented in the overall conclusions (Sect. 7.6). Since these small industries are supplied mainly from the drinking water network of the Water and Wastewater Company there might be overlaps with the water balance of urban water use.

The water extraction of industries is presented with a certain volume per year (m^3/a) and for the different water sources: surface water, groundwater and water supplied by the Water and Wastewater Company.

Current industrial water consumption is presented and discussed with data from the reference year 2006 (and 2012 for agricultural industry). 2006 was a normal and wet year and due to water scarcity in the basin and limited economic development[1], consumption patterns of industry have not changed significantly over the past 10 years. Therefore, data from the year 2006 are presented in the following as current consumption patterns.

Furthermore, different trend and development scenarios are presented and discussed as an outlook for industrial water consumption in the year 2025.

[1]The international trade embargo on Iran was an important factor for a limited industrial development in the country.

7.3 Large Industrial Units

In close collaboration with the Isfahan Water Board Company the 30 largest industries were selected to be included in this study. The selection criterion for large industrial units was a current minimum water consumption of 500,000 m^3/a.

The analysis of the selected industries was based on provided data and communication with the Iranian institutions as well as quantitative surveys and interviews with specific industries. Quantitative surveys and interviews with selected industries contributed to a better understanding of water and wastewater management practices of the industrial units. Based on our interviews we learned that industrial waste water volumes and disposal are very complex and unexplored topics and require further investigation. Water management in large industries only partly follows international standards. Water efficiency is rather high and parts of the produced wastewater are evaporated in open ponds or used for onsite green space irrigation. Nevertheless further research can pave the way to more efficient water use and internal wastewater reuse, as well as possibilities of using treated urban wastewater in industry or reusing industrial wastewater in agriculture.

7.3.1 Current Water Consumption (2006)

The current surface and groundwater consumption of large industries were estimated based on water licenses of industries for the year 2006 transmitted by the Water Board Company.

During interviews with representatives of the largest industrial water consumers, it was found that three companies, Mobarakeh Steel Company, Esfahan Steel Company and Polyacryl Iran, combined use approx. 24 MCM/a less water than permitted by their licenses. In contrast, with 12 MCM, the consumption of the Power Plant Islamabad is twice as high as their water license. The compiled consumption patterns for the selected 30 large industries are formatted in yearly water extraction and presented in Fig. 7.1.

All selected large industries combined consume approximately 114 MCM freshwater per year. Four of these industries are named "other industries". These virtual industries have been defined in collaboration with the Isfahan Water Board Company in order to represent unaccounted for large water users in the catchment.

Drinking water supply for workers of industrial units is already included in the calculations of the urban water sector, hence not considered in the industrial water balance and not further discussed in this report.

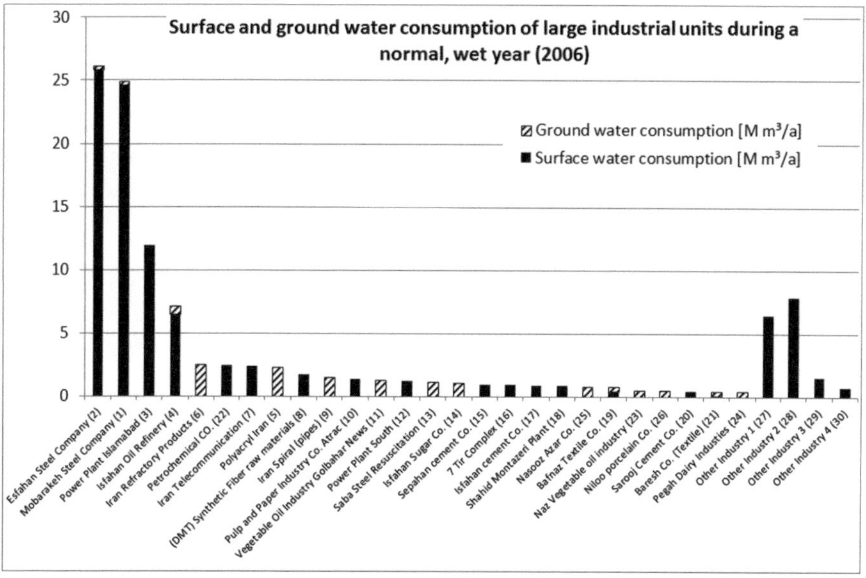

Fig. 7.1 Large industrial units in the catchment area with reference number in brackets

7.3.2 *Current Water Sources (2006)*

Different water sources were identified as important information for a water management process:

- Surface water
- Extracted directly from the Zayandeh Rud river;
- Extracted from one of the irrigation canal networks;
- Extracted from shallow wells close to the river, defined as bank filtration;
- Groundwater from industrial wells
- Water supplied by the Isfahan Water and Wastewater Company.

The share of surface water and groundwater extraction is specified by the water licenses of the specific industries (see Fig. 7.1). In 2006, large industries used about 100 MCM of surface water and 14 MCM of groundwater.

In collaboration with the Isfahan Water Board Company and the Water and Wastewater Company, it was found that during dry years, some large industries (Mobarakeh Steel Company, Esfahan Steel Company and Polyacryl Iran, Power Plant Islamabad, Isfahan Oil Refinery, Shahid Montazeri Plant, Petrochemcial Co. and Pegah Dairy Industries) compensate a temporal lack of surface water access by water supplies of around 6 MCM/a from the Water and Wastewater Company. The overall water consumption of all industries is constant during wet and dry years.

7.3.3 Future Water Consumption in 2025

The future trend of industrial water consumption is important for the meaningful planning of water resources. However, future industrial water consumption highly depends on global and regional socio-economic and political developments. Vice versa, the availability and development of regional water resources and management of the relevant sectors are the limiting factor for possibilities to settle new, water intense industries in the future.

As a result of scarce water resources in the catchment, new industrial development is currently limited. The establishment of new industries is only permitted exceptionally and requires a complex administrative process between Industrial Organization and the Water Board Company, and has to be approved by a regional committee of authorities (Committee # 24).

In the course of the IWRM Zayandeh Rud research project and through constant communication and exchange with the Iranian partners, two different scenarios for future water consumption of large single industrial units in the year 2025 were set up.[2] The two scenarios are largely similar but comprise some differences which result in a 107 MCM/a variance in the estimated future water consumption of large industries.

The first (and in view of the authors more likely) scenario is based on the fact that water resources are continuously decreasing and that water scarcity will have its respective impact on industrial water management decisions. The outlook expects a comparably moderate expansion of existing water consumption and settlement of new large industries, due to limited freshwater resources and moderate economic growth in the region. It includes the hypothesis that the ambiguous development goal for the industrial sector in terms of water consumption will not be fully implemented, but to a limited extent only. Scenario 1 anticipates that a production increase of large industry is granted rather by increased water efficiency and on-site wastewater reuse than on an expansion of their water licenses and consumption.

The second scenario is based on regional and national development plans, if these were to be fully implemented. The outlook assumes a stronger increase of water consumption of some existing industries as well as the establishment of additional new large industries.

7.3.3.1 Scenario 1(Allowing for Decreasing Water Resources)

For the water extraction outlook for 2025, the research team assumes that at most four additional large industries will settle in the Zayandeh Rud basin and influence

[2]While decisions on small and medium industrial settlements are rather taken at provincial level, the location of large industries is a national and strategic matter. For this reason the scenario development was only facilitated for this industrial segment and with the participation of national representatives.

the current water balance of the catchment. Existing extraction licenses will rise only for power plants and one refinery by 2025. The following four aspects are included in the scenario:

- During interviews with the industrial organization it was found that it is very likely that four new large steel and petrochemical industries are to be developed in the eastern part of the catchment due to development planning of the central government. These industries are expected to consume a total of 30 MCM/a in 2025, but their location and sources of water extraction remain unknown;
- The three companies mentioned in Sect. 7.3.1 (Mobarakeh Steel Company, Esfahan Steel Company and Polyacryl Iran) that currently use 24 MCM/a less water than their license permits them, will reach extraction rates up to full license capacity in 2025;
- Due to current limited production capacity, the Isfahan Oil Refinery will be expanded significantly. This will more than double the freshwater extraction with an increase of approx. 17 MCM/a;
- Due to population growth and increased industrial activities it is expected that all power plants will increase their water consumption by 10% by 2025. This also applies to the Power Plant Islamabad with its current overconsumption. The Shahid Montazeri Plant is directly connected to the Isfahan Oil Refinery and will increase its capacity and water consumption by almost 60% to 1.5 MCM/a, equivalent to the expansion of the oil refinery. All development factors will increase the water consumption of power plants by 2 MCM/a in total.

In scenario 1 the annual water consumption of large industry will rise by more than 60% to 187 MCM/a by the year 2025.

7.3.3.2 Scenario 2 (Full Implementation of National Development Plans)

For the water extraction forecast for 2025, it is assumed that most of the existing extraction licenses will not be expanded until 2025. Nevertheless the following increases in water extraction and new developments are expected:

- 12 new large industrial units will be developed within the Zayandeh Rud catchment according to development goals, with a total annual water consumption of 98 MCM. The focus of new developments will be on the steel and petrochemical and chemical sector. The specific location and water sources of these industries are still unknown;
- The three companies mentioned in Sect. 7.3.1 (Mobarakeh Steel Company, Esfahan Steel Company and Polyacryl Iran) that currently use 24 MCM/a less water than allowed according to their licenses will reach an extraction volume up to full license capacity in 2025. Furthermore, the two steel works will increase water consumption beyond that and combined will extract 28 MCM/a more than permitted according to their current water licenses;

- Due to current limited production capacity, the Isfahan Oil Refinery will be expanded significantly. This will more than double the freshwater extraction with an increase of approx. 20 MCM, which is 3 MCM/a more than expected in Scenario 1;
- Due to population growth and its direct connection to the Isfahan Oil Refinery and the Said Montazeri Plant, the power plants Islamabad and Sahid Montazeri Plant are expected to be expanded by 2025 and increase their combined water consumption by 11 MCM.
- In Scenario 2 the annual water consumption of large industry will rise by 160% to 294 MCM/a by the year 2025.

7.3.4 Interim Conclusions

Currently the total freshwater withdrawal of the selected large industrial units is 114 MCM.

Figure 7.2 shows the location and current water consumption of large industrial units in four classes, from less than 1 MCM/a to more than 10 MCM. The map displays the locations of the large industrial water consumers (grey circles) with their reference number (red number) from Fig. 7.2. The map shows that the majority of water extraction takes place upstream of Isfahan City along the Zayandeh Rud

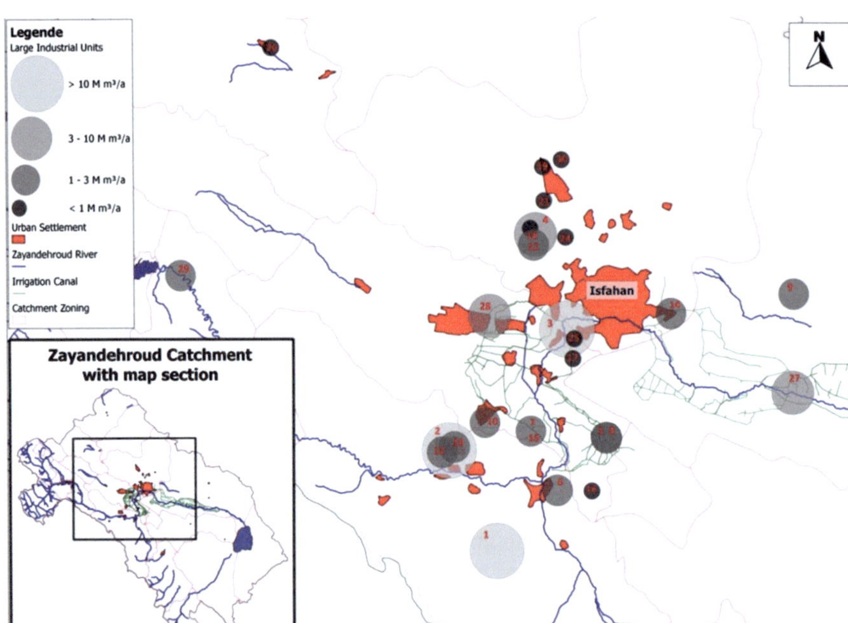

Fig. 7.2 Location and classes of current water consumption (*grey circles*) of large industrial water consumers (*red reference number* from Fig. 7.1) in the catchment

Table 7.1 Current and future water consumption of industrial sectors according to development scenarios

Sector of water consumption	Steel	(Petro) Chemical	Power	Building material	Building material
Current consumption (2006)	MCM	54	17	14	3
	% of total water extraction (114 MCM)	47%	15%	12%	3%
Future consumption (2025) Scenario 1	MCM	89	53	16	3
	% of total water extraction (187 MCM)	48%	28%	9%	2%
Future consumption (2025) Scenario 2	MCM	151	89	25	3
	% of total water extraction (294 MCM)	51%	30%	9%	1%

River or through irrigation networks which have their headwork there. It should be noted that the Borkahr irrigation system, which is located north-west of Isfahan city, is not on the map.

Table 7.1 shows that currently the dominant industrial segments of the large individual industries are steel, petrochemical and chemical industries as well as power plants. The four largest water consumers (Mobarakeh and Esfahan Steel Companies, Power Plant Islamabad and Isfahan Oil Refinery) consume 70 MCM/a (see Fig. 7.1), which is more than 60% of the total water consumption of large industries.

Furthermore the future trend scenarios presented in Sect. 7.3.3 anticipate a further expansion and increase of water consumption of the already dominant sectors. In Table 7.1 the sectoral increase in water consumption is presented for both scenarios outlined above. The calculation for these figures was based on the assumption that the industries which will be developed in the future are either steel or petro-chemical or chemical industries. Therefore the water consumption of these sectors rises extensively in both future scenarios.

The map in Fig. 7.3 shows which industrial units are expected to grow by 2025 according to Scenario 1. The displayed green number is the expected growth of water consumption by 2025 of the industrial unit (in percent, grey circle) where the number is written on top. The map shows that few large industries that extract water upstream of Isfahan city are expected to grow. The Isfahan Oil Refinery as well as the connected Shahid Montazeri power plant north west of Isfahan extract water directly from the Zayandeh Rud upstream of Isfahan. In scenario 2 mainly the same industrial units will grow, but to a higher degree (see Sect. 7.3.3). New industrial developments of both scenarios could not be located on the map.

Regarding the sources of water, currently (2006) only nine large industrial units rely mainly on groundwater and 21 companies rely mainly on surface water. The total groundwater consumption of large industries is only 14 MCM, in contrast to the surface water consumption of 100 MCM/a (88% of total consumption). Freshwater is mainly extracted directly from the Zayandeh Rud River (see Fig. 7.4).

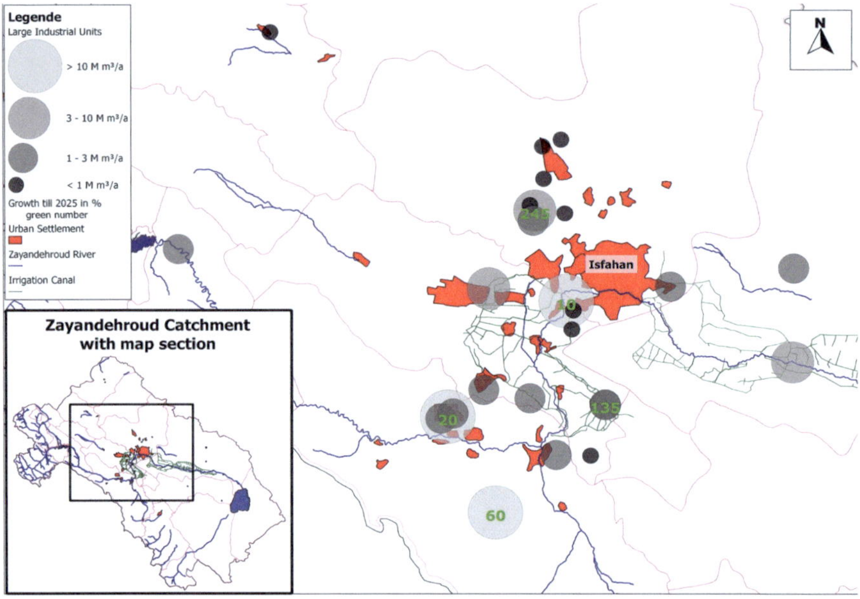

Fig. 7.3 Expected growth in water consumption of units (*green number* as %) on the map of location and current water consumption of large industrial units (Scenario 1)

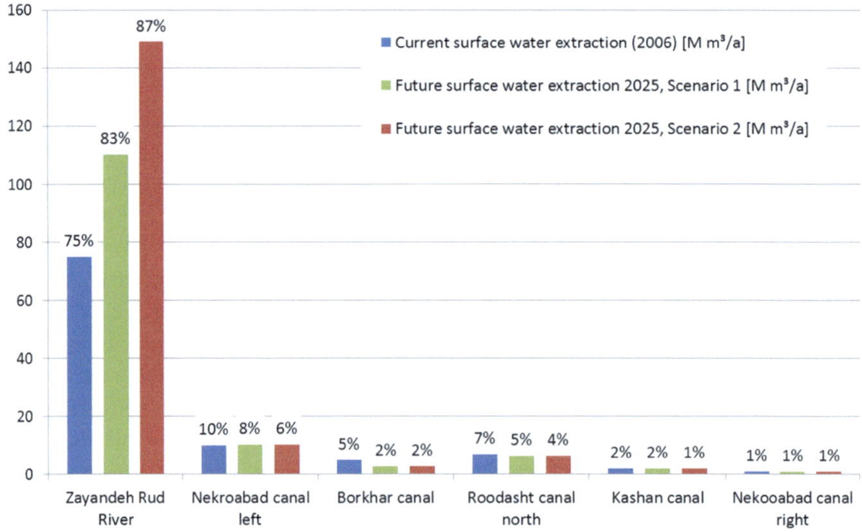

Fig. 7.4 Sources of surface water extraction in 2006 and in the future in 2025 according to scenarios 1 and 2, with percentage of total surface water extraction of large industry

In Scenario 1, the total water withdrawal of large industries is expected to grow up to 187 MCM/a by 2025. The four new industrial developments in the eastern part of the catchment (see Sect. 7.3.3) are expected to have a total consumption of about 30 MCM, while the water source is unknown and therefore not included in the following figure. For the industries that have already been established, the water sources are expected to develop towards higher surface water use. In the forecasted water withdrawal scenario for 2025, groundwater extraction is expected to stagnate at approximately 14 MCM, and water supplied by the Water and Wastewater Company will rise to 10 MCM. The surface water extraction is expected to grow to 133 MCM/a (without the new developments) mainly due to increased extraction by large industrial units directly from the river (see Fig. 7.4).

Analogue, the expected surface water extraction of large industry in 2025 according to Scenario 2 is presented in Fig. 7.4. The planned 12 additional industries with approx. 98 MCM, water consumption are not included in the figures due to unknown water source and location. Besides the 172 MCM/a (without the new developments) expected surface water extraction, the amount of groundwater and water supply by the Water and Wastewater company is equal to Scenario 1.

Presented figures indicate that the high water consumption of large industries, particularly of the steel and petrochemical segment in the central part of the basin, are expected to grow even further in the near future.

7.4 Industrial Settlements and Zones

There are industrial settlements and industrial zones in the catchment. In the past, there have only been industrial zones, which are industrial units, clustered in one area. Industrial settlements have several industrial units in the same area, but also have a common administration in terms of a management board and are handled as one legal entity in terms of distribution of water licenses etc. In the future, the entire industrial development will be organized within industrial settlements; the only exceptions are very large or heavy polluting units. For simplification, in the subsequent document, the term industrial settlement will be used for settlements as well as for zones.

In close collaboration with the Isfahan Water Board Company, Industrial Organization and Industrial Settlements Organization, the 29 largest and most important industrial settlements with at least 10.000 single units were selected.[3] The entities either already have high water consumption or are expected to grow intensively by 2025, reaching a minimum water consumption of approximately 500,000 m^3/a. Currently the 29 settlements together consume approximately 28 MCM.

The locations of the industrial settlements were communicated by the Water Board Company. The Mahmoodabad and Jarghuyeh settlement is, geographically, located a few hundred meters outside the catchment, but since it is supplied by the Water and Wastewater Company, it is included in the water balance of the Zayandeh Rud catchment.

[3]The following zones and settlements were selected: Khomeinishahr, Mahmoodabad, Dolatabad, Morchehkhort, Faridan, Se Rahe Mobarakeh, Tiran, Karvan, Esfidvajan, Oshtorjan, Najafabad1, Najafabad2, Jey, Segzi, Kohpayeh, Harand, Varzaneh, Ezhieh, Meimeh, Alavijeh, Dehagh, Large north Isfahan settlement, Komshecheh, Mohamadabad, Jarghuyeh, Chadegan, Poodeh, Tudashk and Montazeriah.

Following the suggestion of the Water Board Company we clustered the following neighbouring settlements into five compound industrial settlements: Greater Area of Najafabad (Tiran, Karvan, Esfidvajan, Oshtorjan, Najafabad1 and Najafabad2), Greater Area of Jey (Jey and Segzi), Greater Area of Harand (Kohpayeh, Harand, Varzaneh and Ezhieh), Alavijeh and Dehagh, as well as Mahmoodabad and Jarghuyeh settlement. This approach results in 18 settlements in total, which are presented in the following.

Due to missing data, no quantitative information on wastewater production or management could be included in the study. However, it was communicated by the Industrial Settlement Organization that six industrial settlements: Mahmoodabad, Se Rahe Mobarakeh, Oshtorjan, Najafabad1, Alavijeh and Morchehkhort, have a wastewater treatment plant installed, and constructions for a treatment plant are ongoing in Jey industrial settlement. By 2025, 17 industrial settlements are supposed to be equipped with a wastewater treatment plant. Effluent from wastewater treatment plants is intended to be used for green space irrigation and in Morchehkhort settlement 50% of treated wastewater is being reused in the production process. To better understand these situations and their progressive approaches, the aspect of wastewater management in industrial settlements requires further investigation. In order to reach higher water efficiency in industrial settlements, detailed research on different industrial branches and the possibilities of wastewater reuse between industrial units should be elaborated on to reduce water withdrawal and the amount of polluted water emitted into the sewage system and open water bodies. Currently a case study on wastewater reuse between selected industrial units of the Morchehkhort settlement according to the Eco Industrial Park concept is being prepared by the IWRM research project.

7.4.1 Current Water Consumption (2006)

The water demand of each settlement is calculated according to a calculation formula (Formula 1) used and communicated by the Isfahan Industrial Settlement Organization. The formula is based on the built-up area of the industrial settlement in hectares. This way an estimated water demand of the settlement with the currently built-up area is calculated.

Formula 1

$$\text{current water demand of settlement} \left[\frac{\mathrm{L}}{\mathrm{s}}\right] = \text{build surface of settlement [ha]} \times 0.3$$

This calculation is applied for each settlement without considering the branches and other criteria for water consumption of the industries in the settlement. Using

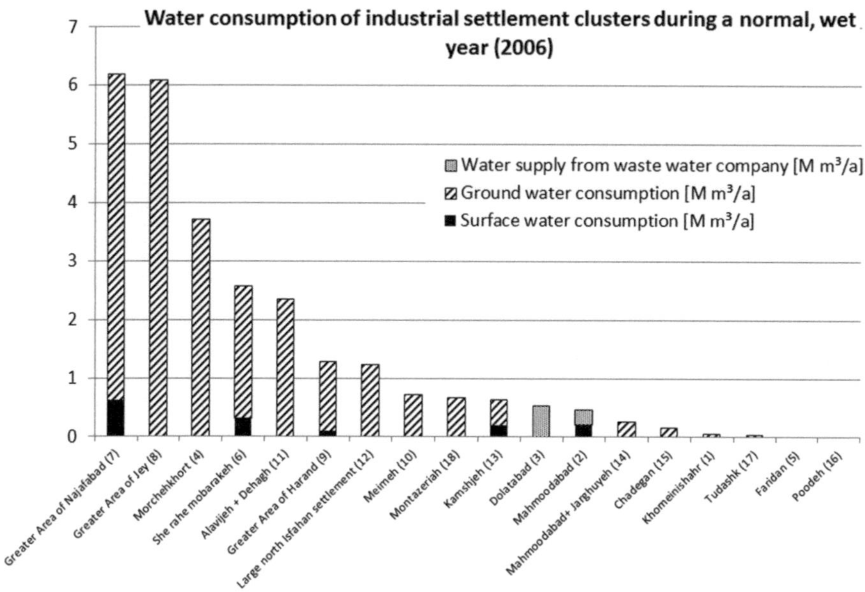

Fig. 7.5 Industrial settlement clusters with water consumption (2006)

this approach, the current (2006) water consumption of the 18 settlements was calculated and is displayed in Fig. 7.5.

7.4.2 Current Water Sources (2006)

The surface water consumption of industrial settlements is based on surface water licenses received from the Isfahan Water Board Company. Figures for water supplied by urban water[4] have been communicated by the Isfahan Water and Wastewater Company. The groundwater extraction of each settlement is calculated according to Formula 2 by adding the supplied surface water and urban water and subtracting it from the calculated water demand of the settlement.

Formula 2

$$\text{Ground water extraction} = \text{water demand (Formula 1)} - (\text{supplied surface water} + \text{urban water})$$

Analogue to large industrial units, the sources of water are:

[4]The water is usually treated and diverted from the urban drinking water supply network and as such is called urban water.

- Groundwater extraction from wells located in the area of the industrial settlements;
- Surface water extraction from the river and canals which are close to the industrial settlements;
- Water supplied by the Water and Wastewater Company, which was initially extracted at the Chamaseman Dam for drinking;
- The Alavijeh and Dehagh settlement is supplied by the Kashan Canal which has its headwork at the main Zayandeh Rud dam and is accounted for as surface water.

During dry years, the Water and Wastewater Company provides water to more industrial settlements than in normal wet years. The overall water consumption of all industrial settlements is constant in dry and wet years.

7.4.3 Future Water Consumption in 2025

Future water consumption in 2025 was forecasted by taking the following approach. For each settlement there exists a total settlement area and a currently built-up area. The assumption is that by 2025 all industrial settlements will develop their full surface and reach maximum water extraction. With the total settlement area and Formula 3, the future water demand of settlements can be calculated.

Formula 3

$$\text{future water demand of settlement}\left[\frac{\text{L}}{\text{s}}\right] = \text{total settlement area [ha]} \times 0.3$$

The establishment of new industrial settlements is currently restricted by the administration, due to water scarcity in the catchment. Requests from investors for new water licenses posed to the Water Board Company are currently pending.

7.4.4 Interim Conclusions

Figure 7.6 shows a map section of the Zayandeh Rud catchment with the locations of industrial settlements including their water consumption (grey circles) and their reference number (red number) from Fig. 7.5. The map shows that the settlements are loosely scattered throughout the catchment and are not as concentrated along the river as large industrial units. Compared to the large industrial units presented in Fig. 7.2, the total consumption of each displayed industrial settlement (grey circle) is smaller.

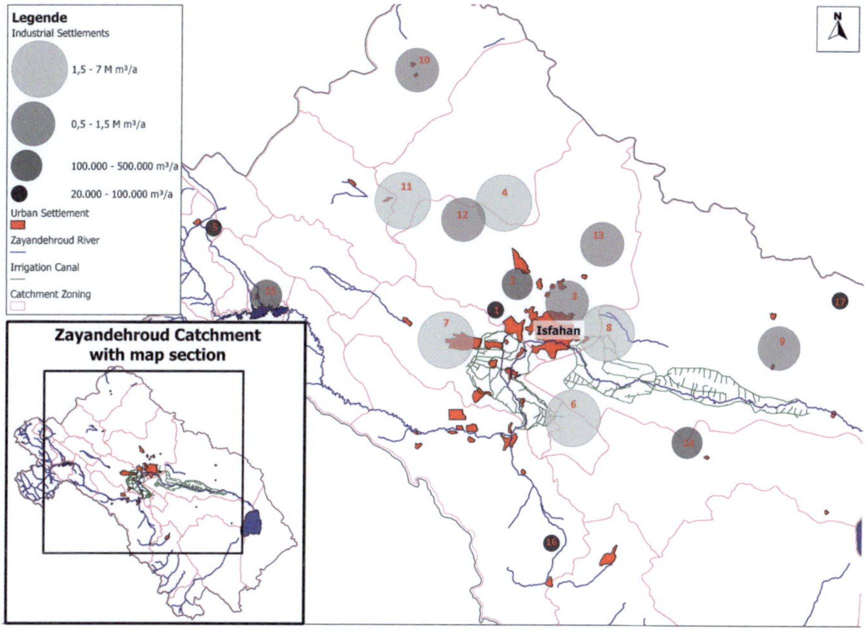

Fig. 7.6 Location and classes of current (2006) water consumption (grey circles) of industrial settlements (with red reference number) in the catchment

Currently (2006) the water consumption of industrial settlements is 27 MCM, of which about 92% (25 MCM) originates from groundwater. By 2025 all industrial settlements are expected to be fully built-up which will more than double the total water consumption to 62 MCM.

The map in Fig. 7.7 shows the industrial settlements that are expected to grow by 2025. The green figure represents the expected percentage growth of water consumption of the industrial settlement (grey circle) by 2025 where the number is written on top. The map indicates that particularly small settlements scattered throughout the catchment are expected to grow excessively by 2025.

In 2025 groundwater is expected to account for 83% (51 MCM) of the total water consumption of industrial settlements.

The presented figures show that the water consumption of industrial settlements will grow intensively in the next few years, putting pressure particularly on groundwater resources. If wastewater treatment capacities of industrial settlements do not grow as predicted, increased water consumption is expected to also impact negatively on the quality of surface water bodies.

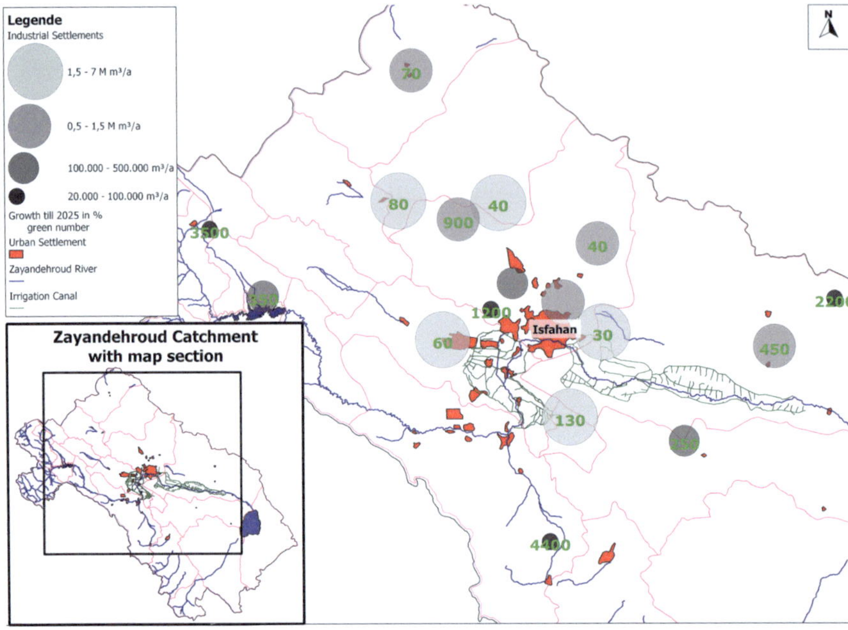

Fig. 7.7 Expected growth in water consumption of industrial settlements (green number in %)

7.5 Industrial Agriculture

Large agricultural production sites have been identified by the Jihad Agricultural Organization and it was decided by the Water Board Company to include them in the industry sector due to their industrial water consumption patterns. Specifically poultry farms, livestock farms (cattle, calves, sheep, goats and camels), greenhouses (vegetables, nursery of flowers and ornament plants) as well as aquaculture production sites (cold and thermal water fish) are defined as industrial agriculture and presented in this section.

The data originate from the Jihad Agricultural Organization and were provided by the Water Board Company.

7.5.1 Current Water Consumption

In contrast to the previously presented data, no specific units or exact locations of agricultural production sites could be presented due to a lack of data. Nevertheless, water consumption in the reference year 2012 is assigned to the counties in the Zayandeh Rud catchment in Table 7.2.

According to the Isfahan Water Board Company, consumption data are based on existing water licenses and county based statistics. No information on the amount of

Table 7.2 Current (2012) water consumption of industrial agriculture by county (source: Jihad Agricultural Organization)

Aquaculture	Greenhouses	Livestock	Poultry	Location
Total water demand (m3/a), 1391	Total water demand (m3/a), 1391	Total water demand (m3/a), 1391	Total water demand (m3/a), 1391	County
7,764,163	2,300,900	2,921,958	1,038,632	Isfahan
346,896	190,000	925,122	207,324	Borkhar
324,821	562,250	579,487	432,655	Tiran&karvan
368,971	14,100	558,188	46,969	Chadegan
63,072	1,264,700	641,445	125,005	Khomeynishahr
1,103,760	600,000	756,197	658,241	Shahinshahr& Meymeh
0	591,000	776,105	466,336	Shahreza
151,373	20,000	479,942	108,617	Faridan
3,815,856	25,650	116,272	22,794	Freydoonshahr
372,125	6,385,800	1,076,282	210,656	Falavarjan
387,893	27,000	211,799	121,585	Lenjan
977,616	2,423,000	597,481	114,776	Mobarakeh
1,056,456	383,260	1,163,503	738,694	Najafabad
0	1,333,500	0	0	Samirom (dehaghan)
16,733,002	16,121,160	10,803,780	4,292,283	

agricultural units or wastewater production could be delivered. Nevertheless, it is assumed that the single units consume less than 500,000 m^3/a each, and are dispersed widely throughout the basin.

7.5.2 Current Water Sources

No data regarding water sources or geographic locations of extraction points could be transmitted. According to the Isfahan Agricultural Organization, all industrial agriculture units cover their water demand by groundwater extraction.

7.5.3 Future Water Consumption in 2025

For the water extraction forecast for 2025, the Isfahan Water Board Company communicated data on the expected percentage growth of the different types of industrial agriculture. The predicted consumption is presented in the conclusions.

Table 7.3 Current and future water consumption of industrial agriculture (source: Isfahan Water Board Company)

	Aquaculture	Greenhouses	Livestock	Poultry
Current water consumption (2012) in MCM	16.7	16.1	10.8	4.3
Future water consumption (2025) in MCM	30.0	30.4	19.3	7.7

7.5.4 Interim Conclusions

Currently (2012), the overall water consumption of different sectors of industrial agriculture is 48 MCM. According to the Water Board Company, water consumption will rise by 55% to 87 MCM/a in 2025 (Table 7.3).

The current and expected future water consumption of industrial agriculture is significant. In particular aquaculture and greenhouses have very high consumption patterns and should be localized and analysed in detail.

7.6 Overall Conclusions and Outlook

The more than 13,000 industrial units analysed (reference year 2006) currently consume a total of approximately 200 MCM/a freshwater. Industrial water consumption is expected to increase by more than 70% to 347 MCM/a in 2025 according to future Scenario 1. In the future, according to Scenario 2, industrial water consumption rises by 130% due to excessive growth of large single industries to 454 MCM. In this calculation, water demand of new large industries which are expected to be settled in both future scenarios was added to the surface water extraction (see Table 7.4).

Currently, 50% of water originates from surface water, 45% from groundwater and 5% from urban water supplied by the Isfahan Water and Wastewater Company. In the future scenarios the proportion of urban water is expected to grow slightly, and in Scenario 2 the surface water proportion will grow by 10%, reducing the share of groundwater.

The expected growth of industrial water consumption in the future is a serious challenge for water management in the Zayandeh Rud catchment. All new or expanding water users in the closed basin will curtail available water resources of other users. The focus of development should therefore lie on a modern, water efficient industry with high economic return and low water consumption, as well as wastewater reuse.

114 MCM or 57% of the total 200 MCM of total industrial water consumption, are accounted for by the 30 largest industrial water consumers. The four largest water consumers (Mobarakeh and Esfahan Steel Company, Power Plant Islamabad and Isfahan Oil Refinery) alone consume - with a share of 70 MCM/a - around 35% of the total industrial water used in the basin. With few large production sites, steel

Table 7.4 Summary of industrial water consumption in the Zayandeh Rud catchment

Summary of industrial water users in the basin	Withdrawal (2025)			Withdrawal (dry year)			Withdrawal (normal & wet year)		
	WW (m^3/a)	GW (m^3/a)	SW (m^3/a)	WW (m^3/a)	GW (m^3/a)	SW (m^3/a)	WW (m^3/a)	GW (m^3/a)	SW (m^3/a)
Single Industrial Units (Scenario 1)	10,530,000	13,855,000	163,000,000	6,140,000	13,960,000	93,980,000	0	14,090,000	99,980,000
		187,000,000			*114,000,000*			*114,000,000*	
Single Industrial Units (Scenario 2)	10,530,000	13,860,000	269,930,000						
		294,000,000							
Industrial Settlements (Scenario 1+2)	9,100,000	51,650,000	1,240,000	5,400,000	21,690,000	0	800,000	24,840,000	1,450,000
		62,000,000			*27,000,000*			*27,000,000*	
Small Units is Isfahan municipality (Scenario 1+2)	10,000,000	0	0	10,000,000	0	0	10,000,000	0	0
		10,000,000			*10,000,000*			*10,000,000*	
Industrial Agriculture (Scenario 1+2)	0	87,350,000	0	0	47,940,000	0	0	47,940,000	0
		87,000,000			*48,000,000*			*48,000,000*	
Summed up consumption (Scenario 1)	29,630,000	152,855,000	164,240,000	21,540,000	83,590,000	93,980,000	10,800,000	86,870,000	101,430,000
		347,000,000			*199,000,000*			*199,000,000*	
Summed up consumption (Scenario 2)	29,630,000	152,860,000	271,170,000						
		454,000,000							

industry is the dominant water consumer in the catchment and sustains a large number of industrial units as suppliers, sub-contractors or subsequent processers. Steel and petrochemical industries are most likely going to expand their dominance in water consumption in the future (see Table 7.1). The total share of the largest industrial water users of total industrial water consumption is - with 55% according to Scenario 1 and 65% according to Scenario 2 – not expected to change or only to grow slightly by 2025 (Table 7.4).

The IWRM Zayandeh Rud project focuses its ongoing research on the reduction of freshwater consumption of large single industries. By applying results of a systematic flow analysis of water, energy and other consumed resources in large industries, processes are intended to be interwoven more efficiently, reducing the overall water and energy demand of industries while preserving productivity. The German research team collaborates closely with one of the large steel works to present state of the art approaches to develop solutions for more water efficient industrial production and internal reuse of process water. As a second pillar, options for reusing treated municipal wastewater as process water in large industries are assessed by the research project, aiming to reduce the overall freshwater consumption in the basin.

With 14%, the approximately 10,000 small industrial units within settlements do not have a high share of the current overall water consumption. Also in the future (2025), small industries are expected to increase their share by up to only 18%, even though their water demand is expected to more than double. However, the small industries' share of groundwater extraction, currently at 45% and around 55% in the future, can be considered significant. Furthermore, wastewater from industrial settlements is not currently, nor will it be in the near future, treated comprehensively. The impact of wastewater discharge is a threat to the quality of water resources and a technical problem for municipal wastewater treatment plants if released into the sewer.

The IWRM project addresses these issues with a research focus on more efficient water management within industrial settlements. The goal is the reduction of freshwater demand and wastewater emission of industrial settlements and as a result reducing groundwater extraction and wastewater emissions. The concept of Eco Industrial Parks is the guiding principle to improve collaboration and coordination amongst small industrial units within industrial settlements to improve water efficiency by trading and reusing process water flows amongst units.

The small industrial units within Isfahan municipality are expected to receive constant water consumption supplied by the Isfahan Water and Wastewater Company and are not part of further research.

Industrial agriculture is, with a current and future share of 24% of the total industrial water withdrawal, a relevant industrial water user. In comparison to agricultural water consumption in the basin, though, industrial agriculture has only a minor share of the water consumption of this sector.

It is important to gain a deeper understanding of the water use and wastewater production of industrial agriculture. More detailed analysis and monitoring of water consumption patterns and water productivity as well as a spatial localization of water users is required. Currently, one important component of the IWRM research project is the analysis of the water productivity of different agricultural practices and the evaluation of alternatives to traditional agriculture in the basin.

References

Bakhtiari S, Iravani M, Hosseini D (2002) The structure of industry in Isfahan Province. Iran J Trade Stud (IJTS) 6(24):7

Ghasemian M, Poursafa P, Amin M, Ziarati M, Ghoddousi H, Momeni SA, Rezaei AM (2012) Environmental impact assessment of the industrial estate development plan with the geographical information system and matrix methods. J Environ Public Health 2012:407162. doi:10.1155/2012/407162

Reisi M, Aye L, Soffianian A (2011) Industrial site selection by GIS in Isfahan, Iran. In: 19th International Conference on Geo informatics

Yekom Consulting Engineers Company (2013a) Report of water use and need in industry and mining and produced wastewater until summer of 2006 in Gavkhuni basin, studies of the country water master plan in salt lake, Gavkhuni, Siahkuh-Rig Zarrin and Kavir Markazi, vol 8. Deputy of Water and Wastewater, Ministry of Energy, Tehran

Yekom Consulting Engineers Company (2013b) Report of water use and demand and produced wastewater in industry and mining until 2041, studies of the country water master plan in salt lake, Gavkhuni, Siahkuh-Rig Zarrin and Kavir Markazi, vol 20. Deputy of Water and Wastewater, Ministry of Energy, Tehran

Zayandab Consulting Engineers Company (2009) Studies of water resources and water use in Zayandeh Rud Catchment, vol 9. Industry, Isfahan Water Board Company, Ministry of Energy, Tehran

Chapter 8
Water Use for Drinking Water and Reuse of Treated Wastewater

Jörn Heppeler

8.1 Introduction

The average annual precipitation is 130 mm in the Zayandeh Rud catchment. Comparing the precipitation values with the average monthly temperatures in the range from 3 to 29°C, the Zayandeh Rud catchment is an arid region. Considering such climatic conditions, it is no surprise that the predominant water withdrawal is taken from the groundwater (57%) and only 43% from surface water (FAO 2015).

The United Nations Environmental Program (UNEP) classified all countries according to their freshwater availability. Thereby the availability of 0 to 1000 m^3 per capita and year is considered "scarcity", from 1000 to 1700 "stress" and 1700 to 2500 "vulnerability" (UNEP 2008). The actual renewable water re-sources in Iran provide 1946 m^3 per capita and year (FAO 2015); hence the country is classified as vulnerable. On top of that water resources are distributed unevenly over the country.

In Iran the predominant amount of water is consumed by the agricultural sector. In 2004 the agricultural water withdrawal amounted to more than 92% whereas more than 50% are gained from groundwater sources. Industrial water consumption is only responsible for slightly more than 1% of the water abstraction while the domestic water withdrawal sums up to more than 6% (FAO 2015). Due to the high water demand and the varying water quality requirements, the agricultural sector can be a suitable consumer of TSE. Subsequently the valuable groundwater resources could be preserved.

J. Heppeler (✉)
p2m berlin GmbH, Berlin, Germany
e-mail: joern.heppeler@p2mberlin.de

S. Mohajeri, L. Horlemann (eds.), *Reviving the Dying Giant*,
DOI 10.1007/978-3-319-54922-4_8

8.2 Water Use

Before TSE is evaluated and recommendations for possible areas of applications are identified, the private water consumption should be looked at. In the following chapter the findings of the private water consumption patterns that have been investigated by water meters in 30 households in Isfahan are presented. The investigation intends to optimize the water consumption and in doing so save valuable fresh water resources.

According to the WHO/UNICEF Joint Monitoring Programme (JMP) for Water Supply and Sanitation close to 95% of the Iranian urban population had access to drinking water sources via house connections in 2011. The figure is 9% lower for the rural population (JMP 2015). The urban population enjoys close to complete coverage of drinking water services. In 2006 the coverage amounted to 99% (FAO 2015).

In Isfahan the drinking water supply is mainly based on surface water. About 80% of the drinking water is withdrawn from rivers or surface water reservoirs. The remaining 20% originate from groundwater sources (Cornel 2005).

Despite the scarcity of fresh water resources there is little information on the water consumption pattern. Further research is required to eventually address target-oriented recommendations for water saving measures.

To overcome this shortcome the water consumption of 30 private households in Isfahan was investigated at various domestic locations. Water meters were installed for two weeks in all of the households. After evaluation of the results, the data sets of 10 households turned out to be complete and reasonable. Accordingly, these data were summarized. The results are given in Tables 8.1 and 8.2.

The total renewable water resources in Iran are different from those in Germany. In Iran 137 billion m^3 and in Germany 154 billion m^3 are available (FAO 2015). In this regard the comparison cannot be seen as best-practice study. The comparison of the average consumption figures of Isfahan to the German values is not adequate due to the data collection in only 10 households in Isfahan. The measurement campaign in Isfahan has a considerable error rate since about 11% of the consumed water could not be attributed.

Therefore, for a comprehensive assessment of the Iranian water consumption pattern further investigation is necessary.

Despite the comparison shortcomings the figures indicate the following profound differences or imprecisions. In spite of limited fresh water resources in Iran, in Isfahan the daily per capita water consumption is about 20 litres higher than in Germany. Still, on a global scale both water consumptions are relatively low.

Table 8.2 shows that the water consumption for showering and bathing are relatively similar. However, there are differences for the toilet flushing. Against the background of exclusive toilet paper usage in Germany, the water consumption for toilet flushing is much higher in Germany. The toilet paper use in conjunction with an increasing spread of water saving technology through push plates in

Table 8.1 Per capita water consumption in Isfahan (Iran) and Germany. Source for German values (BDEW 2015)

	Isfahan–Data of 10 households (randomly chosen, average values) recorded in 2014	Average values of German households of 2014
Daily drinking water consumption per capita [litre]	140	121

Table 8.2 Water consumption of private households in Isfahan (Iran) and Germany

	Isfahan–Data of 10 households (randomly chosen, average values) recorded in 2014	Average values of German households of 2014
Bath room [%]	24	36
Sink/body care [%]	8	
Toilet flushing [%]	16	27
Kitchen (cooking, drinking) [%]	28	4
Laundry [%]	8	12
Dish washer [%]	5	6
Cleaning, gardening, car cleaning [%]	no data	6
Small business [%]	no data	9
Not attributable [%]	11	–
Sum [%]	100	100

Source for German values (BDEW 2015)

Germany gives rise to expectations of lower values than in Iran. Here no explanation can be given; this means further research is necessary.

The water consumption in Isfahan's kitchen is by far higher than in Germany. In Iran homemade food and family gatherings enjoy a high relevance; this might be the reason for a more abundant usage.

This rough comparison illustrates that the water consumption pattern of private households in Isfahan and Germany is similar for the major part of the consumption figures. The only exception is the water consumption in the kitchen. Here the Iranian values are about ten times higher.

Due to the water quantity and quality requirements for safeguarding the livelihood there are various options for reducing the consumption in the kitchen; among them are: repairing trickling water-taps, installation of water-saving fittings, washing of fruits and vegetables in bowls, complete filling of washing machine and dishwasher, etc.

8.3 Reuse of Treated Wastewater in Agriculture

8.3.1 General Conditions in Iran

As outlined above Iran faces an increasing pressure on water resources. In order to tackle this, reuse of TSE can substitute water withdrawal from surface or groundwater. The Iranian Expediency Council supports this measure by the following strategies and policies (Mahmoudian 2004):

- Exploitation of drinking water potential of freshwater prior to its use for other purposes, and allocation of the produced wastewater to the agricultural sector after treatment;
- Replacement of agricultural water rights for withdrawal of freshwater with water rights for TSE in order to ensure future urban water demand;
- Irrigation of green spaces using TSE instead of freshwater being a resource for potable water production;
- Increase of pressure on industries to implement wastewater treatment and recycling facilities (threat of withdrawal of licenses for water abstraction or water supply); and
- Promotion of research projects for the development of reasonable standards for safe and reliable water reuse for agricultural purposes and artificial groundwater recharge (this includes convincing the farmers of advantages regarding the replacement of freshwater by TSE).

In addition to the substitution of freshwater resources, reuse of TSE for agricultural or landscape irrigation provides further benefits: TSE is a reliable perennial water resource containing valuable nutrients that are essential for plant growth, so that application of TSE may also contribute to increased agricultural cost efficiency due to enhanced agricultural productivity and reduced use of chemical fertilizers.

Certain quality standards for the safe application of TSE reuse practices have to be implemented for the protection of field workers and consumers. In case hygienic parameters are exceeded the application of TSE reuse constitutes a serious threat.

In the past Iran hesitated with the bulk usage of TSE. This is underlined by the figures of FAO's database AQUASTAT; in 2003 only 154 million m^3 of TSE were reused (FAO 2015). Until today the figures are only slowly catching up. Tajrishy published figures for reused wastewater of 2010. Therein the reuse is given with 328 million m^3 which is about 40% of the treated wastewater (see Table 8.3).

However, the official figures do not represent the situation on site. The withdrawal of treated effluents for irrigation purposes is widely spread. Aerial photos underline this assumption as in the vicinity of the wastewater treatment plants' (WWTP) effluent channel the area is cultivated and abundantly covered with vegetation. Therefore it seems likely that the unofficial figures for TSE reuse are much higher.

Nevertheless, the TSE reuse quantity in Iran - having more than 78 million inhabitants - is minor. One reason is that in 2010 less than half of Iran's population was connected to the sewer system. In ancient times wastewater was frequently

Table 8.3 Summary of wastewater discharge, collected, treated and reused in Iran in 2010 (Tajrishy 2011)

	In Iran in 2010 [million m^3]
Wastewater produced	3,547
Wastewater collected	1,162
Wastewater treated	820
Wastewater reused	328

Table 8.4 Cumulative wastewater collection network length, number of wastewater treatment plants (WWTPs) and population connected statistics in Iran (Tajrishy 2011)

Year	Length [km]	No. of WWTPs	Population connected
1997	9,978	30	1,959,548
2000	15,654	37	2,327,702
2005	23,473	84	6,001,322
2010	35,500	129	12,977,079

discharged into existent Qanat Systems (also called kareez) or wells (Angelakis et al. 2012). Only in Isfahan the aforementioned method could not be applied and a sewer system was built.

In terms of the sewer system and the wastewater treatment capacity there is room for improvement. The nationwide development over the last years is shown in Table 8.4.

8.3.2 Wastewater Infrastructure in Isfahan

Isfahan the third largest city in Iran is located at the Central Plateau at the edge of the Zagros mountain range. The seasonally fluctuating river Zayandeh Rud flows through the city centre. Due to the geological conditions the city was the first one in Iran to be equipped with a modern sewerage system and WWTPs. The implementation of such investment in the infrastructure was done in the late 1960s and early 1970s (Mohajeri and Dierich 2011).

Today nearly all inhabitants are connected to the sewer system in Isfahan; the wastewater production is given as 575,000 m^3/d (Tabatabaei and Najafi 2009). In the latest census 1,756,126 inhabitants were counted (Statistical Centre of Iran 2012). Originally the sewer system was designed as combined system but due to the low precipitation rate, the storm water proportion is virtually irrelevant (Cornel 2005). The urban sewer system discharges the predominant share of wastewater to one of the four treatment plants of Isfahan.

Isfahan's main four treatment plants are (design capacity as population equivalent – PE):

- WWTP Isfahan North (Phase I: 400,000 PE + Phase II: 800,000 PE)
- WWTP Isfahan South (800,000 PE)

Fig. 8.1 View from the digesters of WWTP Isfahan North

- WWTP Isfahan East (250,000 PE)
- WWTP Shahin Shahr (260,000 PE)

The Isfahan Water and Wastewater Company (IWWC) is responsible for the collection and treatment of wastewater, whereas the TSE reuse for irrigation purposes is within the Esfahan Regional Water Board's (ESRW) field of responsibility.

The project work concentrated on the WWTP Isfahan North (see Fig. 8.1). Therefore the following investigation on TSE reuse is based on these effluents. The plant's treatment aim is carbon removal only.

The first construction phase of WWTP Isfahan North, commissioned in 1987, has a design capacity of 400,000 PE corresponding to about 70,000 m^3/d (Cornel 2005). In recent years, the average hydraulic load was 64,000 m^3/d.

Following the first treatment step consisting of screening, grit separation, and primary clarification in two circular clarifiers, there is a biological stage with two separate aeration tanks. At the time of the project implementation the tanks were equipped with rotary brush aerators. In the meantime the aeration system has been changed to subsurface disk diffusers. For the secondary clarification there are two sedimentation tanks. Finally there is a chlorine contact tank, however, no operational and controlled disinfection stage. Subsequently the TSE is discharged into a concrete channel outside of the WWTP that leads to the Zayandeh Rud.

The excess sludge is thickened, digested and dewatered and eventually used in agriculture.

The second construction phase, commissioned in 2008, has a design capacity of 800,000 PE. In recent years the average hydraulic load amounted to 70,000 m^3/d.

The treatment is carried out based on a two-stage activated sludge process (A–B process). The first treatment step consists of coarse screening, grit separation and fine screening. The biological stage consists of two rectangular tanks of stage A being aerated with centrifugal surface aerators, then four intermediate clarifiers preventing the release of A-sludge and stage B with two rectangular tanks equipped with centrifugal surface aerators. The final clarification takes place in four sedimentation tanks. Before TSE is released into the aforementioned concrete channel the supernatant undergoes disinfection but the disinfection management is weak. On a temporary basis bleach solution (sodium hypochlorite) is added. A proper control mechanism is not in place.

The excess sludge is thickened, digested and dewatered and eventually used in agriculture.

The central conclusion of the process analysis underlines the fact that there is a high degree of optimization potential for the treatment process. For instance, the mixed liquor suspended solids (MLSS) and the sludge age are too high, the removal of floatable matter is unsatisfactory and the sedimentation rate of the clarifiers are insufficient thus there is still a relatively high concentration of total suspended solids (TSS) in the TSE. The average data of monthly mean concentrations 2010–2011 amount to 118 mg TSS/l in the first phase. Similarly the BOD_5 removal of the first phase is insufficient; the average data of monthly mean concentrations 2010–2011 is 104 mg/l.

The TSE values of the second phase point in the same direction. The average data of monthly mean concentrations of 2012 was measured with 93 mg TSS/l and the BOD_5 in the effluent showed 82 mg/l as average data of monthly mean concentrations of 2012.

As conclusion, the TSE of both phases contains relatively high TSS and BOD_5 concentrations, which represent a significant constraint in terms of effluent reuse. Besides the loss of biomass due to insufficiently separated activated sludge from the liquid phase, increased TSS concentrations affect the efficiency of disinfection processes due to possible shielding of microorganisms by particulate matter.

For the efficient operation of disinfection processes, TSS concentrations of secondary effluent should be constantly kept below 20 mg TSS/l, and preferably lower (DWA-M 205 2013); (Metcalf and Eddy Inc 2003) – the lower, the better the disinfection efficiency. With regard to effluent disinfection by UV radiation, it has been shown that even a significant increase in radiation intensity does not improve the disinfection efficiency in case of increased TSS concentrations, i.e. >20 mg TSS/l (Metcalf and Eddy Inc 2003). Concerning chemical disinfection processes, abundance of particles also results in significantly increased demands in terms of the disinfectant dosage due to quenching of disinfectants by particles. Thus, suspended solids do not only reduce the disinfection efficiency, but also increase the disinfectant consumption. Moreover, high concentrations of organic matter result in the increased formation of harmful disinfection by-products.

In addition to the impact on disinfection processes, high TSS concentrations may lead to deposits or blockage of certain irrigation systems (e.g. drip irrigation and subsurface irrigation). Therefore, the reliable reduction of the TSS concentrations of TSE generated at WWTP Isfahan North is of major importance with regard to any agricultural reuse scheme.

In large WWTPs being appropriately designed and operated, TSS concentrations can commonly be kept below the target value (20 mg TSS/l). In case of increased and/or significantly varying TSS concentrations, additional filtration of secondary effluent may be required in order to ensure sufficiently low TSS contents prior to disinfection. However, the dimensioning of filtration processes is also strongly depending on the TSS concentrations of the effluent to be filtered. Therefore, it is strongly recommended to focus on the minimisation and stabilisation of the secondary effluent's TSS concentration by operational measures prior to designing any tertiary filtration stage for the WWTP Isfahan North.

8.3.3 Iranian Limits for Wastewater Reuse

Despite the aforementioned advantages that have been recognized by the Iranian government, the country lacks a policy for wastewater reuse. In 1994 the "Effluent Discharge Standard" was developed by the Department of the Environment (Bahri 2008). The standard is only applicable for discharge of treated wastewater into receiving surface water bodies, into absorption wells and for irrigation purposes in agriculture (Iranian Department of Environment 1998). In this context, surface waters are defined as seasonal or permanent rivers, natural or artificial lakes and lagoons. The term absorption well refers to ditches or trenches having the capacity for the infiltration of water into the ground. The limit values are given in Table 8.5 is. Being limited to these three reuse options, the standard does not represent a complete wastewater reuse policy.

Concerning the key question of the present project of TSE reuse options in agriculture the "Effluent Discharge Standard" forms the basis.

Compliance with the effluent standards specified in Table 8.5 is under the supervision of the Department of Environment. Amongst others, the following specifications associated with these effluent standards are to be considered:

- Measurements for the control of compliance with the standards shall be on the basis of combined samples, i.e. 24-hour samples which have been prepared of individual samples being taken at intervals of maximum 4 h.
- The effluent shall be free of odour, foam, and floating matter. The colour and turbidity of the effluent shall not visibly change the natural appearance of receiving and local water bodies.
- Discharge of TSE into absorption wells or trenches is prohibited if the distance between their base and the groundwater table is less than 3 m.
- Polluting industries must treat their effluents up to a standard level according to engineering principles and the use of an appropriate and economic technology.

Table 8.5 Iranian Standard for discharge and reuse of treated sewage effluent (Iranian Department of Environment 1998)

Parameter	Unit	Discharge into surface waters	Discharge into absorption wells	Reuse for agricultural irrigation
Silver	mg/l	1	0.1	0.1
Aluminium	mg/l	5	5	5
Arsenic	mg/l	0.1	0.1	0.1
Boron	mg/l	2	1	1
Barium	mg/l	5	1	1
Beryllium	mg/l	0.1	1	0.5
Calcium	mg/l	75	–	–
Cadmium	mg/l	0.1	0.1	0.05
Free Chlorine	mg/l	1	1	0.2
Chloride	mg/l	600[a]	600[b]	600
Formaldehyde	mg/l	1	1	1
Phenol	mg/l	1	negligible	1
Cyanide	mg/l	0.5	0.1	0.1
Cobalt	mg/l	1	1	0.05
Chrome(VI)	mg/l	0.5	1	1
Chrome(III)	mg/l	2	2	2
Copper	mg/l	1	1	0.2
Fluoride	mg/l	2.5	2	2
Iron	mg/l	3	3	3
Mercury	mg/l	negligible	negligible	Negligible
Lithium	mg/l	2.5	2.5	2.5
Magnesium	mg/l	100	100	100
Manganese	mg/l	1	1	1
Molybdenum	mg/l	0.01	0.01	0.01
Nickel	mg/l	2	2	2
Ammonia	mg/l	2.5	1	–
Nitrite	mg/l	10	10	–
Nitrate	mg/l	50	10	–
Phosphate-P	mg/l	6	6	–
Lead	mg/l	1	1	1
Selenium	mg/l	1	0.1	0.1
Sulphide	mg/l	3	3	3
Sulphite	mg/l	1	1	1
Sulphate	mg/l	400[c]	400[d]	500
Vanadium	mg/l	0.1	0.1	0.1
Zinc	mg/l	2	2	2
Oil & grease	mg/l	10	10	10
Detergent	mg/l	1.5	0.5	0.5
BOD_5	mg/l	30 (instant 50)	30 (instant 50)	100
COD	mg/l	60 (instant 100)	60 (instant 100)	200

(continued)

Table 8.5 (continued)

Parameter	Unit	Discharge into surface waters	Discharge into absorption wells	Reuse for agricultural irrigation
Dissolved oxygen	mg/l	≥2	–	≥2
Total dissolved solids (TDS)	mg/l	–[a]	–[b]	–
Total suspended solids (TSS)	mg/l	40 (instant 60)	–	100
pH		6.5–8.5	5–9	6–8.5
Radioactive materials	mg/l	0	0	0
Turbidity	NTU	50	–	50
Colour		75	75	75
Temperature	°C	–[e]	–	–
Faecal coliforms	MPN/100 ml	400	400	400
Total coliforms	MPN/100 ml	1000	1000	1000
Nematode eggs	eggs/l	–	–	–[f]

[a]Discharge of amounts above those specified in the table shall be allowed only if the effluent will not affect the concentrations of chloride, sulphate, and dissolved solids in receiving water bodies by more than 10% over a radius of 200 m.
[b]Discharge of amounts above those specified in the table shall be allowed only if the concentrations of chloride, sulphate, and dissolved solids in receiving water bodies will not be increased by more than 10%.
[c]Discharge of amounts above those specified in the table shall be allowed only if the effluent will not affect the concentrations of chloride, sulphate, and dissolved solids in receiving water bodies by more than 10% over a radius of 200 m.
[d]Discharge of amounts above those specified in the table shall be allowed only if the concentrations of chloride, sulphate, and dissolved solids in receiving water bodies will not be increased by more than 10%.
[e]Temperature shall not affect the temperature of the receiving water body by more than 3 °C over a radius of 200 m.
[f]The number of nematode eggs should be ≤1 egg/l if the effluent is to be used for irrigation of crops eaten uncooked.

- Sludge and other solid materials produced by wastewater treatment plants must be treated to an appropriate level before discharge and such an action shall not lead to the pollution of the environment.

In Table 8.5 the parameters faecal and total coliforms as well as nematode eggs refer to microbiological standards. Bacteria such as faecal coliforms and total coliforms present an epidemic risk in case of TSE reuse for irrigational purposes. On a global scale the Iranian limit values are to be seen in the mid-range.

While the occurrence of coliform bacteria gives information about the hygienic properties of TSE, the predominant health risks for consumers and farm workers related to irrigation using TSE are usually associated with intestinal parasites.

These show a substantial resistance against chemical disinfection and have a long resilience under environmental conditions (WHO 2006).

8.3.4 WHO Limit Values for Water Reuse

In the current WHO Guidelines for wastewater use in agriculture an integrated approach combining various options for achieving the water quality required for agricultural irrigation is promoted (WHO 2006). This approach is to some extend in contrast to the Iranian regulation. Besides the production of TSE by more or less technological wastewater treatment processes, the WHO approach includes a differentiation regarding the crops to be irrigated. Moreover, the contribution of different irrigation techniques for the reduction of the risk of crop contamination is considered, as well as appropriate health and safety measures for the protection of field workers and consumers.

With regard to produce restriction, crops which are usually eaten raw are distinguished from crops being only consumed after further processing (e.g. wheat) or cooking (e.g. rice, potatoes) and non-food crops (e.g. cotton, fodder plants). Crops being eaten raw, including vegetable and salad crops (e.g. root crops, lettuce), require irrigation water of high quality in terms of pathogen contents, i.e. water suitable for unrestricted irrigation. TSE having a lower quality should only be reused for restricted irrigation, i.e. for crops being processed before consumption.

However, the above mentioned WHO Guideline follows the concept of health-based targets, which means achieving a certain reduction of health risks related to effluent reuse. As a result of this approach, an overall pathogen reduction of 6 log-units is suggested for unrestricted irrigation of leafy vegetables (e.g. lettuce) and of 7 log-units for unrestricted irrigation of root vegetables being consumed raw (e.g. onions). The reduction of coliform bacteria (e.g. *Escherichia coli*) is considered as a practicable measure for pathogen reduction (WHO 2006).

Whilst the focus of the formerly suggested WHO limits was only on the TSE quality, the now required 6- to 7-log-unit pathogen reduction may be achieved by a combination of different options. This approach is illustrated in Fig. 8.2. For instance, the relatively high requirements for unrestricted irrigation of leafy vegetables are fulfilled in case of conventional irrigation (flood, furrow, or sprinkler irrigation) using treated effluent after 3-log-unit pathogen reduction (being equivalent to 104 thermo tolerant coliforms/100 ml) if the irrigation is stopped 1–2 weeks prior to harvesting and if the produce is sufficiently washed by the consumers (example B in Fig. 8.2). However, the same result can be achieved with a lower degree of wastewater treatment, but effluent application via surface drip irrigation, provided that the crops are high-growing and do not get in contact with the soil (example C in Fig. 8.2). In case of low-growing crops, additional pathogen removal by wastewater treatment prior to surface drip irrigation has to be ensured (example D in Fig. 8.2).

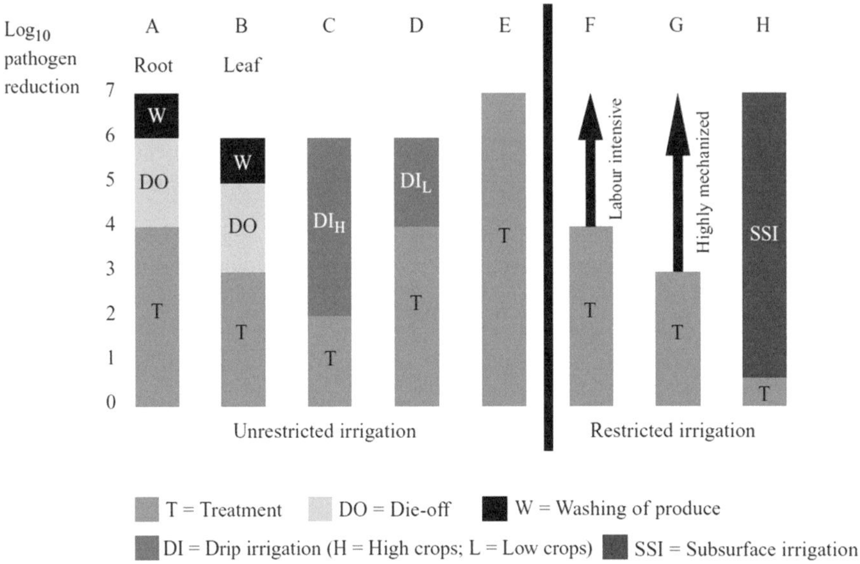

Fig. 8.2 Examples for options for the reduction of viral, bacterial and protozoan pathogens by different combinations of health protection measures (WHO 2006)

In addition to the concentrations of pathogenic bacteria, viruses, and protozoa, which are usually measured by a bacterial indicator parameter (e.g. *E. coli*), helminth eggs contained in effluents are another issue of concern. Regardless of whether the effluent is reused for unrestricted or restricted irrigation, the content of helminth eggs should be <1 egg/l according to the WHO Guideline. In case of possible contact of children with TSE due to playing in the fields, the concentration should be <0.1 egg/l (WHO 2006).

With regard to the Iranian Effluent Discharge Standard, there is no distinction in terms of different effluent qualities for unrestricted and restricted irrigation, except for the limit concerning helminth eggs (<1 egg/l), which is only to be considered for the irrigation of crops being eaten raw (i.e. unrestricted irrigation).

8.3.5 Recommendations for WWTP Isfahan North

Disinfection of the plant's effluent is reasonable and necessary because the TSE deriving from both phases of WWTP Isfahan North is used for agricultural irrigation, so that disinfection contributes to the protection of farm workers and consumers against microbial infections. However, the plant's disinfection conditions are critically assessed for the following reasons:

- The disinfection efficiency is likely to be limited due to high TSS concentrations. Microorganisms being embedded in suspended particles may be shielded against the impact of disinfectants and survive, which may result in bacterial regrowth in TSE.

- High BOD_5 concentrations result in the excessive consumption of free chlorine, thus requiring high chlorine doses for achieving sufficient disinfection efficiency. Moreover, the reaction of chlorine with organic matter results in the formation of potentially harmful disinfection by-products.
- Chlorine-based disinfection is generally less efficient in plants without nitrification due to the formation of chloramines, which provide also some disinfection potential, but require much longer contact times for achieving disinfection efficiencies being comparable to free chlorine.
- There is no management and control strategy in place to supervise the dosage and efficiency. Accordingly there is no information on the effectiveness and possible remaining risks for field workers and/or consumers.
- There are no devices for controlled mixing of the bleach solution into the secondary effluent. Rapid mixing of the disinfectant into the water is highly important in terms of disinfection efficiency.
- Chlorination is not effective against helminth eggs and parasites (especially Cryptosporidium).

Bearing in mind the aforementioned arguments, the most suitable disinfection technology for Isfahan was assessed. The assessment was based on the technologies: maturation ponds, ultraviolet radiation, membrane filtration, chlorination, ozonation, peracetic acid or performic acid. The advantages and disadvantages were carefully weighed and verified for their applicability in Isfahan.

The result is that only a few seem to be appropriate for the implementation at WWTP Isfahan North. Eventually, the disinfection by membrane causes high investment and operation costs. The application of chlorine seems to be not feasible due to the missing nitrification stage. The disinfection by ozonation and peracetic acid forms high risks for the operating staff due to in-situ storage, preparation and application. The recommendable stage for Isfahan is UV disinfection. A clear benefit of the UV disinfection facility is the simple and transparent operation. Furthermore the low labour requirements are favourable, i.e. routine work such as supervision, documentation and cleaning of UV-lamps.

In any case, necessary treatment process improvements at the WWTP Isfahan North in particular for the secondary effluent, need to be implemented before the application is feasible and reasonable. According to latest operational data the permanent achievement of 20 mg TSS/l in the supernatant is highly questionable. Additional treatment stages such as rapid filtration or micro screening have to be introduced. Here the permeate's target value should be 5 mg TSS/l.

Supposing the operation of the WWTP is optimized and the TSE is in line with the Iranian limit values, the reuse in agriculture as irrigation water is a favourable option. But to be well on the way to TSE reuse additional aspects have to be kept in mind. At one side the wastewater is a reliable year-round water resource while irrigation water is required seasonally, so storage becomes an issue which is discussed below. On the other side, the way of irrigation is crucial to the health of the field workers, as outlined above, and has a significant influence on the quality of the final field product.

In Isfahan Province the surface irrigation, including flood and furrow irrigation is the predominant agricultural irrigation technique. In case of hygienically insufficient water quality, the application of surface irrigation involves a high health risk not only for fieldworkers due to direct contact with irrigation water, particularly if protective clothing (e.g., boots, shoes and gloves) is not worn but also to humans living in the irrigated area in general. The latter can suffer from inappropriate constructed and managed irrigation schemes that can provide ideal conditions for the proliferation of bugs that are the basis for vector-borne diseases. Consequently, the population's health is threatened by malaria, bilharzia, dengue and dengue haemorrhagic fever, liver fluke infections, filariasis and onchocerciasis (Feyen and Badji 1993) when the climatic conditions are suitable.

The design of the irrigation scheme has to consider at least the following aspects. In order to minimize the health impacts, straight and lined channels without vegetation lead to faster flows reducing breeding sites. The channels have to be drained in case of no irrigation. Generally covered channels are favourable; however, due to higher construction costs these are often omitted.

The risk for crop contamination with pathogens due to contact with irrigation water is high, in particular for low-growing crops. Therefore, especially flood irrigation implicates increased health risks for consumers if crops being eaten raw are irrigated with insufficiently treated sewage effluent.

Another aspect to be considered is the irrigation efficiency, i.e. the ratio of water needed by the crop to water applied to the field. This ratio is comparatively low in case of surface irrigation systems because of water losses due to evaporation and seepage beyond the root zone. The increase of the irrigation efficiency is considered to be of particular importance in Isfahan due to limited water resources available and a high water demand for agricultural irrigation.

Taking into account the aforementioned aspects, it becomes evident that surface irrigation is not the optimum irrigation technique in Isfahan Province when considering water scarcity and TSE reuse. Here irrigation technics such as drip irrigation are favourable.

Drip irrigation has been widely used in the Middle East for a long time and is not only beneficial regarding irrigation efficiency and salinity control, but also in terms of the reduction of epidemic risks related to effluent reuse. The salt movement is radial along the point of irrigation and salt accumulates between drip points. However, an increased TSE quality regarding suspended solids is required in order to avoid clogging of the emitter system.

Discussing the reuse potential, the storage of TSE is also one important issue. Wastewater is available relatively constantly throughout the year. In contrast thereto irrigation water is required only during the irrigation period. The excess TSE being generated during the non-irrigation period is suggested to be taken into consideration to avoid any significant loss of this valuable water resource.

Storage of TSE in tanks above ground would require the construction of enormous tank capacities and lead to enormous evaporation losses, so that this option is not further considered. However, seasonal aquifer recharge by infiltration of TSE may be an alternative for temporary storage of large quantities of TSE during the non-irrigation period. However, it has to be pointed out that the current treatment process at WWTP Isfahan North including carbon removal only, but no nitrification/denitrification does not fulfil the Iranian standard's requirements on effluent quality for aquifer recharge (cf. Table 8.5).

The comparison of the operational data with the Iranian standard shows that carbon, nitrogen and coliform parameters are well above the limit. Under these conditions the aquifer recharge is not permitted.

Supposing the WWTP is extended by additional treatment steps and the TSE is in line with the Iranian standard's values, the aquifer recharge is a favourable option for groundwater recharge respectively temporary storage of irrigation water during non-irrigation periods i.e. in winter when temperatures are relatively low. Therefore, evaporation and correlated water losses due to large surface areas are considered to be of minor importance for infiltration ponds.

It has to be further considered that infiltration rates may be lower in winter due to the water's increased viscosity. Furthermore a slower drying and recovery of the infiltration capacity has to be anticipated. The biological activity and the associated reduction of the infiltration rates due to bio-clogging of the ponds' bottom may be less intense in winter (Bouwer 2002). To enhance the infiltration capacity injection wells might be considered, however, high TSE quality and intensive maintenance are required in order to avoid clogging.

Detailed knowledge of the hydrogeological conditions at possible infiltration sites and the surrounding areas are mandatory. Surface infiltration requires sufficient vertical soil permeability from the bottom of the infiltration basin to the top of the aquifer. Moreover, detailed information is required in terms of the storage capacities and groundwater flow conditions at the respective sites in order to be able to make reliable predictions regarding the expected dispersion and transport of infiltrated TSE and its influence on natural groundwater resources.

An initial investigation revealed that the area around WWTP Isfahan North belongs to the Borkhar aquifer. According to data provided by the Iranian partners, the electrical conductivity in Borkhar aquifer is in the range of 2000 to 3000 μS/cm, i.e. the groundwater is affected by increased salinity. In the long term, irrigation and/or groundwater recharge using TSE may further affect the quality of soil and groundwater in terms of salinity. Therefore, attention has to be paid to the TSEs salinity and to its effects on soil and groundwater quality, also considering irrigation management. According to today's status the recharge into the Borkhar aquifer has to be seen critically but additional hydrological data is necessary and field tests are required to come to a final recommendation.

8.4 Conclusions

The first section of this chapter compares the private water use in Isfahan to Germany. Obviously the project's result is relatively similar to the German water consumption. Nevertheless, Germany with abundant water resources has a daily per capita consumption around the hundred twenties while Iran with water vulnerability consumes twenty litres more per day and capita. Based on this rough comparison, the implementation of water saving technology in private households is recommended.

In the second section of this chapter the reuse of TSE of the WWTP Isfahan North is investigated and appropriate and safe reuse options in the agricultural sector are discussed. Today, TSE reuse is implemented due to growing scarcity of water resources and increasing water demands. Frequently, the replacement of agricultural water rights for withdrawal of freshwater by water rights for TSE is implemented in order to safeguard future water resources for human consumption.

In the case of WWTP Isfahan Nord the secondary effluent contains high TSS and BOD_5 concentrations which represent a significant constraint in terms of effluent reuse. High TSS concentrations negatively affect the efficiency of disinfection processes due to the shielding of microorganisms by particulate matter against radiation or chemical disinfectants. Provided that the necessary improvements of the secondary effluent are implemented, UV disinfection for Isfahan is the most recommended solution.

Surface irrigation is not the optimum technique in Isfahan Province considering water scarcity and reuse of TSE. A sustainable irrigation management is to be established that e.g. foresees techniques that are adapted to the crop culture with regard to survival time of the pathogens.

It seems to be advisable to consider temporary storage of TSE for the non-growing period. Due to the climatic conditions in Isfahan covered or underground facilities might represent a feasible option.

Based on the previously mentioned conditions it is strongly recommended to optimize the treatment process of the WWTP Isfahan North in order to reduce the BOD_5 and TSS concentrations in the effluent, implement a sustainable disinfection stage and achieve a high quality TSE to be suitable for reuse in agriculture.

References

Angelakis AN, Mays LW, Koutsyiannis D, Mamassis N (2012) In: Jimenez B, Asano T (eds) Evolution of water supply through the Millennia. IWA Publishing, London, UK

Bahri A (2008) In: Jimenez B, Asano T (eds) Water reuse-an international survey of current practice, issues and needs. IWA Publishing, London, UK

BDEW (2015) Bundesverband der Windenergie- und Wasserwirtschaft e.V. Retrieved March 2016, from https://www.bdew.de/internet.nsf/id/3852C5217E9FD4E1C125786C004274E7/$file/Trinkwasserverwendung%20im%20HH%202014-Anteile.pdf

Bouwer H (2002) Artificial recharge of groundwater: hydrogeology and engineering. Hydrogeol J 10:121–142
Cornel P (2005) In: Meßmann S, Schneider T (eds) Anforderungen an die Abwassertechnik in anderen Ländern, vol I. Ruhr-Universität Bochum, Bochum
DWA-M 205 (2013) Disinfection of biologically treated wastewater – German Rules and Standards. Hennef.
FAO (2015) AQUASTAT main database. Retrieved October 2015, from Food and Agriculture Organization of the United Nations: http://www.fao.org/nr/water/aquastat/data/query/index.html?lang=en
Feyen J, Badji M (1993) Environmental and Health Aspects of Irrigation. Zeitschrift für Bewässerungswirtschaft 28:6–30
Iranian Department of Environment (1998) Effluent discharge standards
JMP (2015) Joint monitoring programme (JMP) for water supply and sanitation. Retrieved March 2016, from WHO and UNICEF: http://www.wssinfo.org/documents/?tx_displaycontroller[type]=country_files
Mahmoudian S A (2004) UNESCO - regional center on Urban water manaagement-Tehran. UNESCO Retrieved 14 Oct 2015, from RCUWM: http://www.rcuwm.org.ir/En/Events/Documents/Seminars/Articles/12.PDF
Metcalf & Eddy Inc (2003) In: Tchobanoglous G, Burton FL, Stensel HD (eds) Wastewater engineering–treatment and reuse, 4th edn. McGraw-Hill, New York
Mohajeri S, Dierich A (2011) Der iranische Abwassersektor zwischen Modernisierung und wachsenden Herausforderungen. Korrespondenz Abwasser Abfall 7:662–668
Statistical Centre of Iran (2012) Iran statistical yearbook 2012–2013 (1391).
Tabatabaei SH, Najafi P (2009) Irrigation and Drainage-Effects of irrigation with treated municipal wastewater on soil properties in arid and semi-arid regions, vol 58. John Wiley & Sons Ltd. DOI:10.1002/ird.449
Tajrishy M (2011) Wastewater treatment and reuse in Iran: situation analysis. Environment and Water Research Center (EWRC) – Sharif University of Technology, Department of Civil Engineering, Tehran
UNEP (2008) Freshwater under threat-South Asia. United Nations Environmental Programme, Nairobi
WHO (2006) WHO guidelines for the safe use of wastewater, excreta and greywater-wastewater use in agriculture, vol II. World Health Organization, Geneva

Chapter 9
Development of a Water Quality Monitoring System for the Zayandeh Rud River

Thomas Erhard Balling and Homeira Safigholy

Abbreviations

°C	Degrees celsius
AC	Alternating current
BETX	Aromatic hydrocarbons
BOD_5	Biological oxygen demand in 5 days
CHC	Chlorinated hydrocarbon
COD	Chemical oxygen demand
DC	Direct current
DDT	Dichlorodiphenyltrichloroethane
DO	Dissolved oxygen
EU	European Union
IRWQI	Iran water quality index
NSF	National Sanitary Foundation
PAH	Polycyclic aromatic hydrocarbons
PE	Population equivalent
PHC	Polychlorinated hydrocarbons
SAR	Sodium adsorption ratio
TDS	Total dissolved solids
TOC	Total organic carbon
TS	Total solids
TSS	Total suspended solids
WWS	Wastewater sewer
WWTP	Wastewater treatment plant

T.E. Balling (✉)
KOCKS Consulting Engineers, Koblenz, Germany
e-mail: balling@kocks-ing.de

H. Safigholy
Environmental Department of the Province of Esfahan, Esfahan, Iran
e-mail: Homeira_Safigholy@yahoo.com

S. Mohajeri, L. Horlemann (eds.), *Reviving the Dying Giant*,
DOI 10.1007/978-3-319-54922-4_9

9.1 Introduction

A consortium of scientists and companies collaborated with partners in Iran to develop an implementable integrated water resources management concept in 2013. The Federal Ministry of Education and Research of Germany supported the project. The subject of this cooperation was the central Iranian catchment of the Zayandeh River ("Zayandeh Rud"), an area severely affected by water shortages. The river is vitally important to the economic development of the semi-arid region of Esfahan. The goal was to ensure sustainable water use along the river, taking into consideration the various requirements of agriculture, industry, the residents and the environment (Mohajeri and Nuñez von Voigt 2011; Mohajeri and Kaltofen 2011).

The general picture is that the water quality of the river has become worse during recent decades due to increased wastewater production from the growing population. At the same time the flow of the river has declined as a result of reduced rainfall caused by climate change and increased water demand, in turn reducing the river's self-purification capacity. Nitrate and phosphate concentrations have increased as well as the electrical conductivity of the river water, due to reduced flows and high evaporation especially in the eastern part of the river basin.

More than 15 locations of domestic and industrial wastewater discharge have been recorded along the river. In addition there are numerous unregistered nonpoint locations of wastewater discharge which affect water quality.

The role of KOCKS Consulting Engineers within the project was to develop a monitoring system for determination of the water quality of the Zayandeh Rud, in consultation with both the Environmental Department and the Water Board of the Province of Esfahan. A fact finding mission to Esfahan was carried out in February 2012, including data acquisition, consultation with authorities and on-site inspections of the monitoring and sampling locations along the Zayandeh Rud.

9.2 Existing Situation of the Zayandeh Rud Water Quality Monitoring System

9.2.1 Existing Online and Non-Online Monitoring System

The Environmental Department and the Water Board each maintain their own monitoring stations for supervising river water quality. The stations are indicated in Figs. 9.1 and 9.2.

The Environmental Department operates two online stations (Chamaseman Dam, Sade Tanzimi). These stations are equipped as follows:

- Building (4–8 m^2)
- Sensors: pH, temperature, electrical conductivity, dissolved oxygen, turbidity (Hach-Lange)

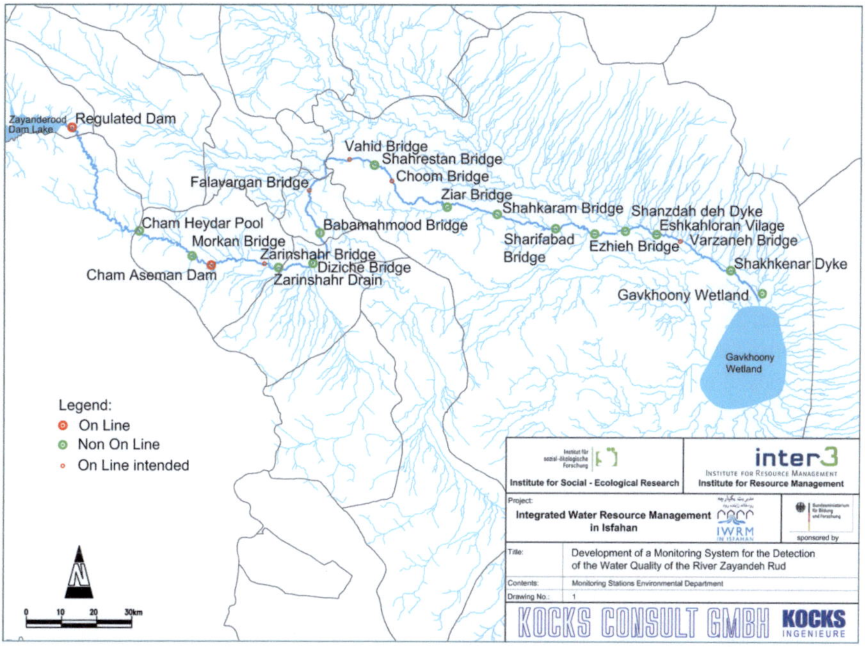

Fig. 9.1 Environmental Department monitoring stations

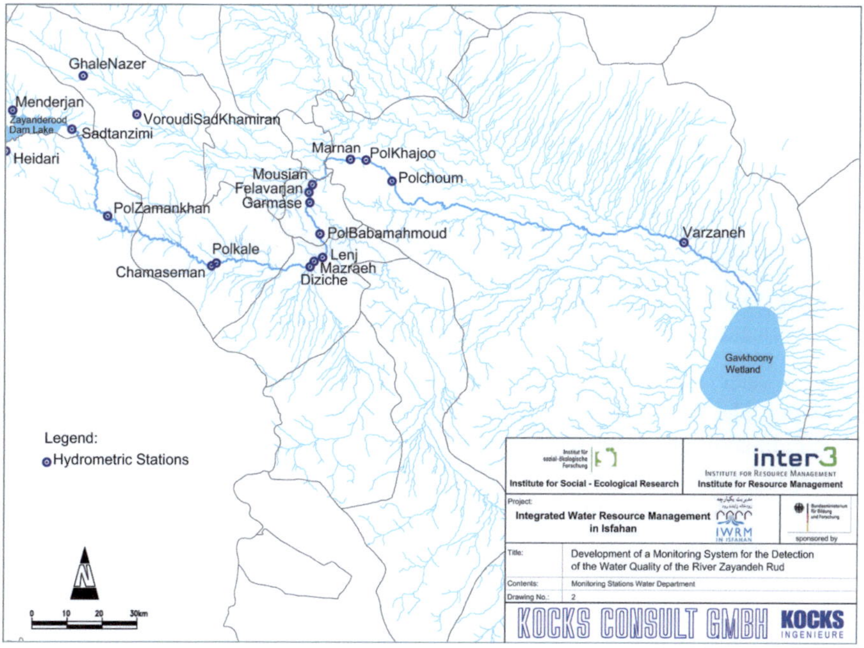

Fig. 9.2 Water Board monitoring stations

- Data logger (Hach-Lange)
- Display module (Hach–Lange)
- Solar panel

The probes are installed and well protected in a rigid steel wire box in the river but close to the banks for easy access. They are wired to a data logger inside the small service building nearby. From here all relevant data are transferred online via telecommunication equipment (modem) to the Environmental Department head office in Esfahan. Maintenance of the probes is carried out at fortnightly intervals and also as-needed. Service of the probe sensors requires wading through knee-deep water since the probes are fixed in position. The probes are taken out of the water during freezing weather, so at this time it is not possible to obtain data online. It is proposed to equip five more stations with similar equipment in order to obtain more online data along the river (Fig. 9.1). The Environmental Department intends to measure the following additional parameters online each hour:

- COD
- NO_3
- PO_4
- SO_4

In addition, water samples are taken eight times per year from 20 stations (both online and non-online stations) for chemical analysis in the laboratory of the Environmental Department (Fig. 9.1). Water is sometimes not available in the dry season at two stations: Shah Karam and Shanzdah Dam.

The Water Board operates 15 stations for hydrometric measurements (Fig. 9.2). They consist of the following components:

- Building (4–6 m^2)
- Steel rope fixed between two masts across the river with a sliding carriage for two persons
- Velocity meter
- Water level gauge
- Chart recorder for water level
- Solar panel
- Electrical data transmitter

The water level of most hydrometric stations is transferred wirelessly to receivers. The flow velocity is measured manually and the flow rate is calculated on site. In addition, water samples are collected regularly at these stations. Chemical analyses are carried out in the Water Department laboratory.

Photographs of existing water quality measurement stations are shown in Figs. 9.3 and 9.4.

Fig. 9.3 Online station Sade Tanzimi

Fig. 9.4 Sensor chamber, online station Chamaseman

9.2.2 *Chemical, Physical and Microbiological Water Analyses*

Chemical, physical, and microbiological analyses for water quality determination are carried out regularly in both the laboratory of the Environmental Department and the laboratory of the Water Board, independently.

9.2.2.1 Technical Equipment of the Laboratories

In both laboratories atomic absorption spectrometry is used for quantitative analyses of chemical elements. Gas and liquid chromatography is used for organic analysis. Both laboratories are comprehensively equipped with the following main apparatus:

- Atomic absorption spectrometer (AAS)
- Gas chromatograph (GC)
- Gas chromatography-mass spectrometry (GC-MS)
- High performance liquid chromatograph (HPLC)
- Ion chromatograph (IC)
- Spectrofluorometer
- Spectrophotometer
- Fume cupboard
- Heated water bath
- Incubator
- Autoclave
- Laboratory centrifuge
- Agar plate
- Magnetic stirrer
- Microscope

The laboratory of the Environmental Department mainly carries out water analyses whereas the laboratory of the Water Department also undertakes sediment analyses.

9.2.2.2 Chemical, Physical and Microbiological Parameters

The following parameters are determined:

Laboratory of the Environmental Department:

Chemical parameters

- Chloride (Cl)
- Fluoride (F)
- Iron (Fe)

- Calcium (Ca)
- Manganese (Mn)
- Nitrate (NO_3)
- Nitrite (NO_2)
- Phosphate (PO_4)
- Ammonium (NH_4)
- Total Dissolved Solids (TDS)
- Total Suspended Solids (TSS)
- Biological Oxygen Demand (BOD_5)
- Chemical Oxygen Demand (COD)
- Total Organic Carbon (TOC)
- Total Hardness
- Alkalinity
- SO_4^{-2}
- S^{-2}

Physical parameters

- Temperature
- Electrical conductivity
- Turbidity
- Dissolved Oxygen (DO)
- pH
- Turbidity

Microbiological parameters

- *Escherichia coli*
- Total coliform

Laboratory of the Water Department:
Chemical parameters

- Chemical Oxygen Demand (COD)
- Biological Oxygen Demand (BOD_5)
- Dissolved Oxygen (DO)
- Nitrate (NO_3)
- Nitrite (NO_2)
- Ammonium (NH_4)
- Aluminium (Al)
- Lead (Pb)
- Tin (Sn)
- Mercury (Ag)
- Iron (Fe)
- Cadmium (Cd)
- Chromium (Cr)
- Copper (Cu)
- Nickel (Ni)

- Phosphate (PO_4)

Physical parameters

- Temperature (°C)
- Turbidity

Microbiological parameters

- *Escherichia coli*
- Total coliform

9.2.3 Determination of the Water Quality

Until 2012 the Environmental Department determined water quality using the NSF Water Quality Index (US National Sanitation Foundation, http://www.water-research.net, Hallock (2002)). This is a 100 point scale that summarises results from a total of the following nine different measurements:

- pH
- DO
- Turbidity
- Faecal coliform
- BOD_5
- Total phosphate
- Nitrate
- TS (total Solids)
- Temperature change

The 100 point index can be divided into several ranges as listed in Table 9.1. Five colours show the quality class of the river in different reaches.

The chemical water quality of the Zayandeh Rud at various reaches in 2011 according to the NSF-Index is shown in Fig. 9.5.

This map should be updated once or twice per year. The most critical part of the Zayandeh Rud is the area east of Esfahan: here water quality is bad due to low absolute flows and lack of constant flow throughout the year. In general water quality becomes worse at locations of wastewater discharge (Chap 9.3.3, Fig 9.10).

The NSF–Index was used until August 2014. It has been replaced by the Iranian Water Quality Index developed by the Environmental Research Institute of Shahid Beheshti University of Teheran. This index includes the following ten parameters:

- BOD_5
- COD
- Dissolved oxygen
- Electrical conductivity
- Faecal coliforms

Table 9.1 Water Quality Index legend according to NSF

Range	Quality	Colour
90–100	Excellent	Blue
70–90	Good	Green
50–70	Medium	Yellow
25–50	Bad	Brown
0–25	Very bad	Red

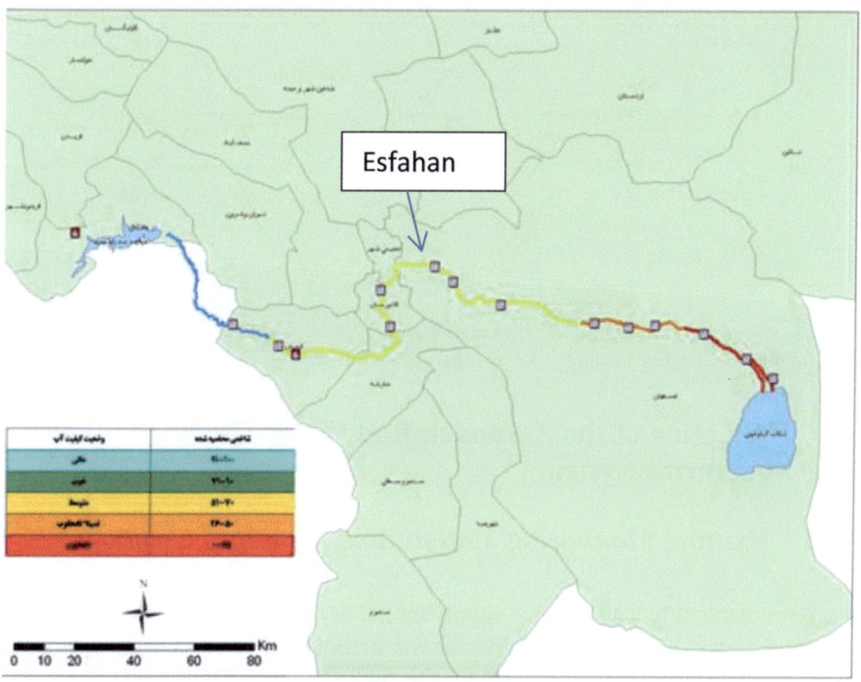

Fig. 9.5 Chemical water quality of Zayandeh Rud in 2011 (NSF–Index, Environmental Department of Esfahan)

- Sodium adsorption ratio (SAR)
- Nitrate
- Total phosphate
- Total hardness
- pH

The 100 point index is divided into seven ranges corresponding to Table 9.2. Seven colours show the quality classes of the river.

The chemical water quality of the Zayandeh Rud at various sections in August and September 2015 according to IRWQI is shown in Fig. 9.6. No water for sampling was available between Zarin Shar and Pole Wahid (grey line).

Table 9.2 Iranian water quality index (IRWQI) legend (Hashemi 2014)

Amount of Index	<15	15-29.9	30-44.9	45-55	55.1-70	70.1-85	>85
Descriptive Equivalent	Very Bad	Bad	Relatively Bad	Medium	Fair	Good	Very Good

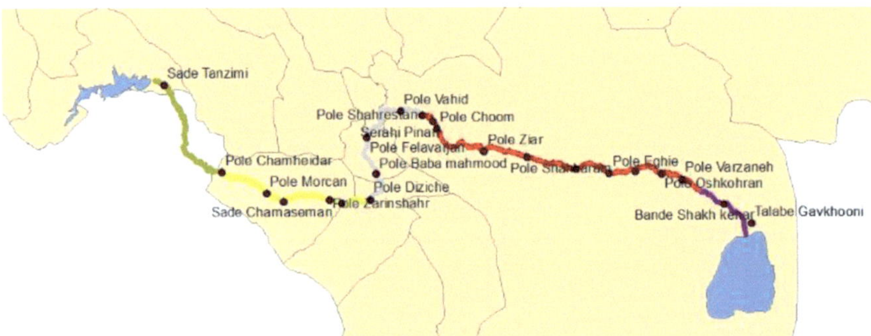

Fig. 9.6 Chemical water quality of Zayandeh Rud in August—September 2015 (IRWQI, Environmental Department of Esfahan)

9.3 Evaluation of the Zayandeh Rud Water Quality Monitoring System

9.3.1 Existing Monitoring Stations and Sampling Locations

The river is the only water source supplying the densely populated Esfahan region (drinking water, industrial water, water for irrigation). There is water pollution along the entire river from just below the Sade Tanzimi Dam to the Gavkhuni Wetlands. There are numerous water quality monitoring stations and sampling locations along the Zayandeh Rud: the Environmental Department runs two online monitoring stations and 18 water sampling locations, and the Water Board operates 17 stations. Most of the stations are downstream of cities and downstream of wastewater discharge points (Chap. 9.3.3). Additional stations are proposed to monitor locations where untreated wastewater is discharged into the river. The existing monitoring stations should be retained for future water quality control since water quality is expected to remain poor at those locations. One online station is located at the drinking water intake at Chamaseman Dam in order to monitor water quality at this location.

Efficiency could be increased through coordination between laboratories. The following stations of the Environmental Department and the Water Board should be combined for chemical analyses to avoid duplication of effort (Fig. 9.7):

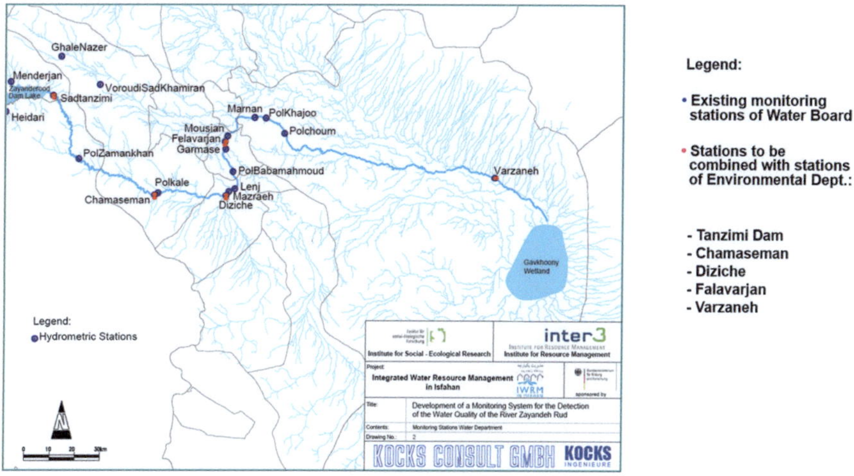

Fig. 9.7 Stations of Environmental Department and Water Board recommended for combination (KOCKS Consult)

- Tanzimi Dam
- Chamaseman
- Diziche
- Falavarjan
- Varzaneh

9.3.2 Proposed Online Monitoring Stations

The following stations are proposed for future online measurement (Fig. 9.1):

- Zarin Shar Bridge (wastewater discharge, city >100,000 PE)
- Falavardjan Bridge (wastewater discharge, city >20,000 PE)
- Vahid Bridge or Chum Bridge (wastewater discharge)
- Varzaneh Bridge (wastewater discharge, city >20,000 PE)

The installation of online measurement at Vahid Bridge (Chum Bridge) depends on the availability of adequate secure space.

Surface water for urban water supplies is abstracted at a rate of about 12 m^3/s at a water intake at the Chamaseman Dam. An oil pipeline crosses the river a few kilometres upstream. There is a risk of contamination of drinking water due to accidents (oil spills). An alarm system connected with the water treatment plant should be considered. The Consultant recommends moving the existing online monitoring station upstream and adding additional equipment for oil spill detection,

Fig. 9.8 Oil detection device in open watercourses (Jola Company, Lambrecht Germany)

such as a floating oil detection device (e.g. JOLA Spezialschalter K. Mattil & Co. KG, Klostergartenstraße 11–20 • 67,466 Lambrecht, Fig. 9.8). This equipment is suitable to detect extensive oil pollution. It should be installed in a low flowing section of the river upstream of the water intake.

The Zarin Shar Station should be moved downstream to the suburban area so as to capture pollution arising from additional wastewater discharge locations.

9.3.2.1 Technical Equipment of Future Online Monitoring Stations

Future online monitoring stations should be equipped with the technical devices listed in Table 9.3.

The sensors for pH, DO, and electrical conductivity listed above also detect temperature. Sensors for online measurement of COD, BOD_5, NO_3 and PO_4 are not recommended for installation since their calibration and especially maintenance requires special training, and also creates much higher investment and maintenance costs. These are certainly higher than the cost of conventional analyses of these parameters in the existing laboratories.

The Consultant suggests fixing the sensors in a float equipped with probe holders to be able to respond well to changing water levels. The minimum water depth is 0.3 m. The float itself should be installed in a perforated stainless steel tube to protect the sensors. A stainless steel chain should be fixed to the float so that it is possible to pull the sensors out of the water for maintenance. The sensor cables should have an appropriate length (Fig. 9.9). Maintenance of this system will be easier since maintaining the existing stations necessitates wading through the water to reach the sensors.

It is proposed to install the proposed online stations on Water Department properties. Security could be achieved by fencing these areas. Power connections will usually not be required since power for measurement and data transfer could be

Table 9.3 Equipment for future online stations and producers (KOCKS Consult)

Parameter or item	Hach-Lange	WTW
pH	1200S SC/LXV400.99.10001	SensoLyt 700 IQ
Temperature	–	–
Turbidity	SOLITAX SC/LXV423.99.10000	VisoTurb 700 IQ
Dissolved oxygen	5740SC/LXV425.99.00001	FDO 700 IQ (optic)
Electrical conductivity	3798S SC/LXV428.99.00001	TetraCon 700 IQ
Data logger	GE SC1000 display module/LXV400.99.10001	MIQ

provided by solar panels. To ensure a more reliable power supply these should be of higher capacity than those currently installed. A water connection is not required: the service team cleans the sensors with water from containers.

9.3.2.2 Maintenance of Online Stations

Maintenance is limited to the following activities:

- Cleaning of floats and sensors
- Calibration of sensors
- Removing the sensors from the water during extreme weather conditions

Maintenance should be implemented twice per month. In case of floods/dry period/freezing additional maintenance is required. To avoid damage, it is of utmost importance to prevent the sensors from becoming dry. This applies to the pH sensor in particular.

9.3.3 Locations of Wastewater Discharge

Locations of wastewater discharge and proposed analyses are indicated in Table 9.4. These locations should be taken into consideration for future investigations (chemical analyses, wastewater research and physical measurement). Basically this list should be completed by competent authorities (Environmental Department and Water Department). According to European Directive 2008/105/EC (European Community 2008) all relevant locations of wastewater discharge should be recorded to assist the development of measures to improve river water quality (See Fig. 9.10).

Flow measuring devices particularly suitable for measuring flow rates are proposed by the Consultant. The combination of a flow profile sensor and an ultrasonic water level sensor would provide the data for a software program to automatically calculate average flow velocity (Fig. 9.11). Mitigation measures to reduce the discharge of wastewater into the river were suggested by the Consultant (Fig. 9.10).

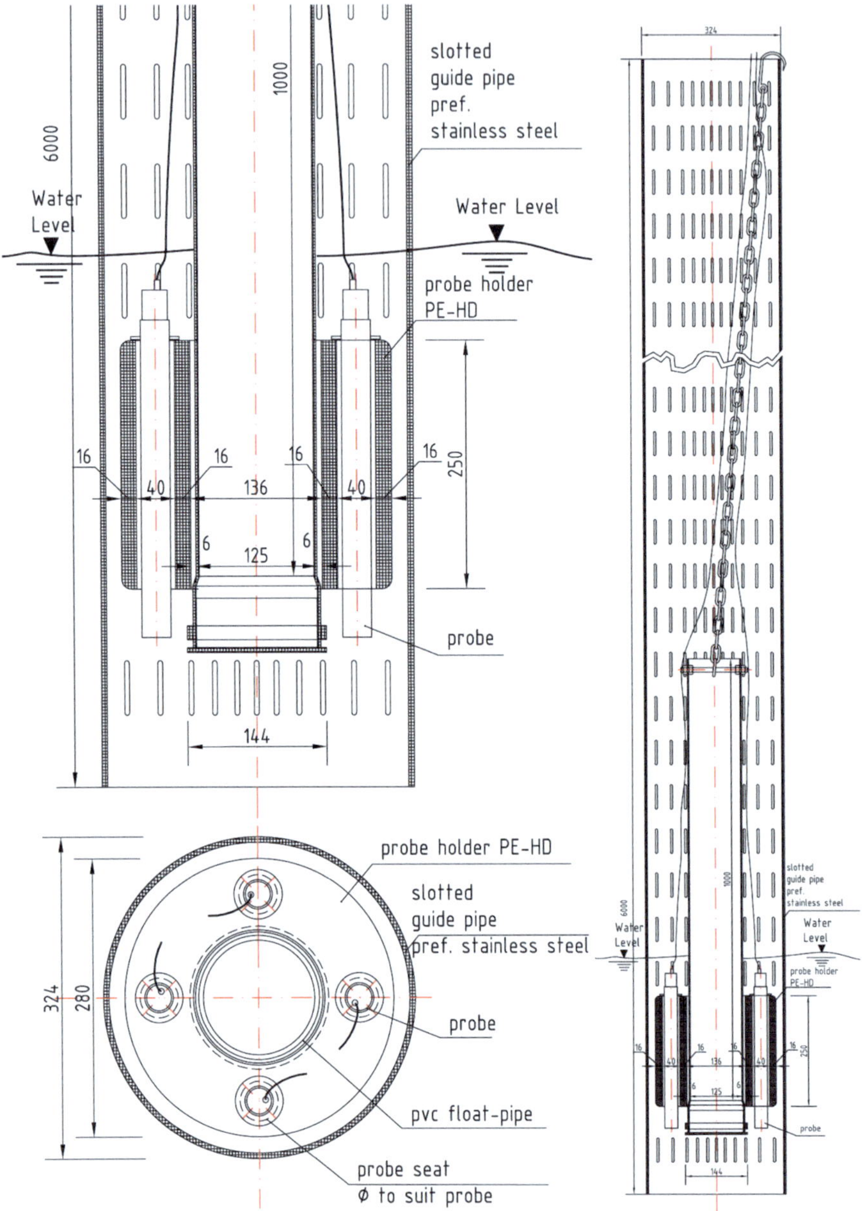

Fig. 9.9 Proposed tube with float and sensors for future online monitoring stations (KOCKS Consult)

Table 9.4 Important locations of wastewater discharge—proposed chemical analyses (KOCKS Consult)

No.	Station name	Coordinates		Domestic waste water	Industrial waste water	Chemical analyses				Saprobic index	Flow measuring, waste water research
		UTM X	UTM Y			BOD5, COD, TSS[a]	Total P, total N[a]	Heavy metals, PAH, CHC[b]	Others		
1	Fish Farm—Sade Tanzimi	480489	3620087			×				×	
2	City of Saman	499382	3591408	Yes		×	×			×	×
3	Shelvan	493034	3593580						Benzene, mineral oil	×	
4	City of Baghbahadoran	518022	3582774	Yes		×	×			×	×
5	Dam Chamaseman	520248	3581771							×	
6	Zarin Shar out-let in channel	539143	3583021	Yes	Yes	×	×	×			×
7	Drain pipe Zarin Shar	539769	3581121	Yes	Yes	×	×	×		×	×
8	Shoor River	548800	3581258	Yes	?	×		×		×	
9	Mobarakeh	550736	3583635	Yes	?	×	×	×		×	×

[a] according EU Directive EU 91/27/EEC
[b] according EU Directive 2008/105/EEC

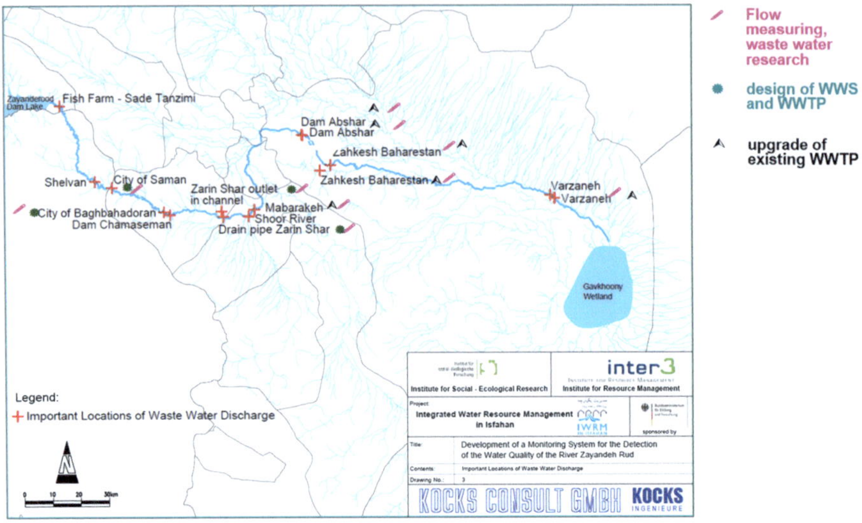

Fig. 9.10 Locations of wastewater discharge—proposed mitigation measures (KOCKS Consult)

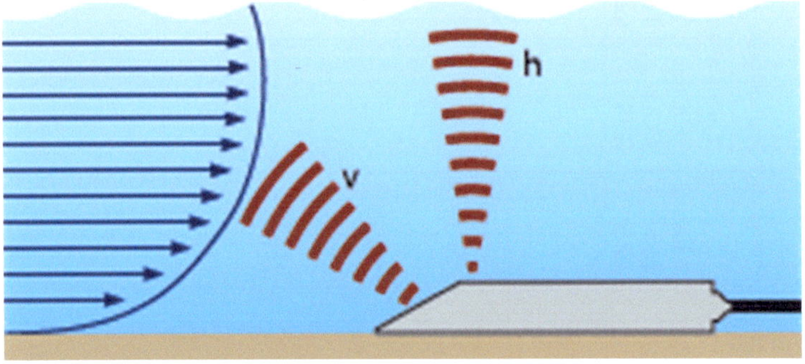

Fig. 9.11 Functional principle of a flow meter measuring mean flow velocity (v) and water level (h) (Wasser—Abwasser—Systemtechnik GmbH, Braunschweig, Germany)

This kind of equipment has proven to be very flexible in the field since it can be operated on a direct current (DC) power supply from a simple battery thus being independent of an external alternating current (AC) power supply. In addition, all data can be stored on a data logger, providing information about the flow rate, the daily discharge and the discharge characteristics (peak flow, low flow).

9.3.4 Parameters for Chemical Analyses

Chemical parameters measured by both laboratories (Environmental Department and Water Department) are listed in Chap. 9.2.2.2. They are sufficient for the determination of the IWQI Index. To cover the entire pollution potential of the river the following additional parameters are suggested (Priority Substances according to EU-Directive 2008/105/EC):

- Heavy metals (nickel, copper, zinc, manganese) and arsenic
- Polycyclic aromatic hydrocarbons (PAH: Benzo(a)-pyrene, Benzo(b)-fluoranthene, Benzo(k)-fluoranthene, Benzo(g, h, i)-perylene, Indeno(1, 2, 3-cd)-pyrene)
- Benzene and derivatives (BETX)
- Chlorinated hydrocarbons (CHC)
- DDT
- Pesticides (e.g. Aldrin, Dieldrin, Endrin, Isodrin and pesticides which are commonly in use in the Esfahan Region)
- Mineral oil hydrocarbons

Chemical analyses should be implemented only if there is evidence for contamination. Pesticide contamination is likely since about 240,000 ha of land is used for agriculture in the vicinity of the river. PAH originating from the air and from the numerous landfill sites and waste dumps alongside the river can also occur in the river water (Fig. 9.12). There is evidence for water pollution with heavy metals, probably discharged by heavy industry in the vicinity of Zarin Shar. Water pollution with mineral oil and benzene can occur in the vicinity of petrol stations, car repair shops and at locations where engine oil is changed. Water pollution with chlorinated hydrocarbons (solvents) is possible from paint shops and car repair shops near the river. In general there is a high risk of contamination of the Zayandeh River by numerous pollutants (Fig. 9.13). Sources of pollution should be recorded and chemical analyses should be carried out accordingly. Parameters for chemical and biological analyses at different locations of wastewater discharge are proposed in Table 9.4 and the locations are shown in Fig. 9.14.

9.3.5 Frequency of Sampling

Samples should be collected once or twice per month for chemical analyses and determination of water quality (NSF-Index, Länderarbeitsgemeinschaft Wasser 1998). However these only provide a snapshot of water quality at the moment of sampling. Therefore shorter sampling intervals are proposed. The EU Directive does not suggest a frequency for sampling: this depends on laboratory capacity and available funds. The capacity and interest of additional laboratories from other institutes should be considered as an option to increase the frequency of sampling and analysis.

Fig. 9.12 Dump site close to the river near the City of Pinart (KOCKS Consult)

Fig. 9.13 Major discharge of wastewater into Zayandeh Rud close to the City of Zakhesh Bakharstan (KOCKS Consult)

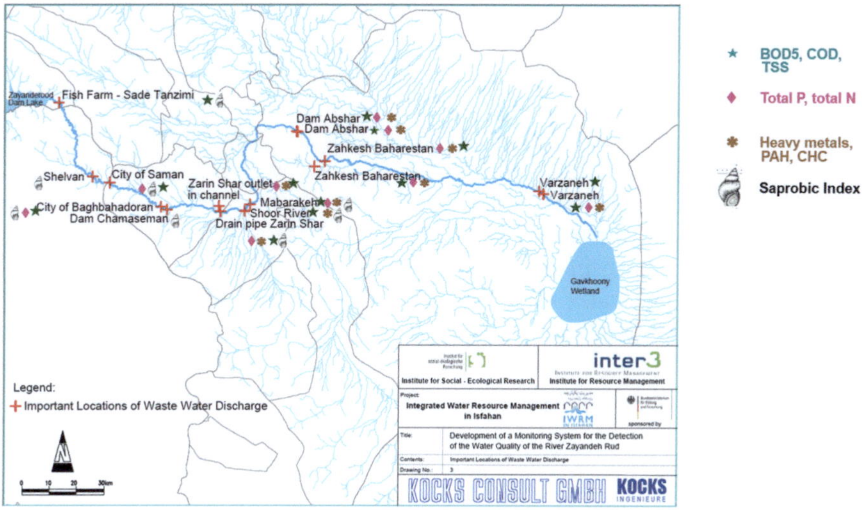

Fig. 9.14 Locations of wastewater discharge—proposed chemical and biological investigations

9.3.6 *Methods for the Assessment of Water Quality*

9.3.6.1 Calculation of Saprobic Index

The saprobic index is correlated with organic pollution and changing oxygen concentrations of running watercourses. Calculation of this index is one component of the assessment of ecological status according to EU Directive 2000/60/EC. The method is described in detail in DIN 38410 (2004).

Different taxa (groups of species) of macroinvertebrates are used as indicator organisms for different classes of organic pollution. The following taxa are abundant in the Zayandeh Rud. They are bioindicators of water quality (Varnosfaderany et al. 2010):

- Insect larvae (Chironomidae, Odonata, Hemiptera, Ephemeroptera, Plecoptera, Trichoptera, Coleoptera)
- Freshwater shrimps (Amphipoda)
- Shells (Bivalvia)
- Snails (Gastropoda)
- Leeches (Hirudinea)
- Worms (Oligochaeta)

Considering the large amount of organic pollution in the river, mostly originating from untreated domestic wastewater, calculation of the saprobic index is recommended.

It would be useful to undertake further studies of this topic since knowledge of the life histories of the benthic organisms of the Zayandeh River is lacking. More taxonomic work is needed to identify the macroinvertebrates to species level to improve the accuracy of the index. The saprobic weight and saprobic value of taxa also have to be determined. Therefore at present the saprobic system cannot be applied. The University of Technology, Department of Natural Resources Esfahan, has already implemented research work concerning a saprobic system for the Zayandeh Rud (Varnosfaderany et al. 2010), and should be involved in future studies.

The best sampling locations on the Zayandeh Rud are upstream of Esfahan since here there is flow in the river throughout the year. It is proposed that existing sampling locations downstream of wastewater discharges upstream of Esfahan are used for the saprobic index (Fig. 9.14). The index should not be calculated for brackish water with high salinity. It is also not useful to calculate the index for locations with no flow, according to DIN 38410 (2004). Consequently most locations downstream of Esfahan cannot be used for calculation of the saprobic index.

9.3.6.2 Chemical Water Quality + Saprobic Index

Determination of chemical water quality (NSF) is an option that should be carried out regularly in future. Since this only provides a snapshot of the current status of the water quality it should be complemented by the saprobic index, which indicates water quality over the past few weeks. It is possible to calculate the NSF at any location of the river whereas determination of the saprobic index makes sense only for reaches of the river with flowing water. Maps showing the chemical and biological water quality of the river at various reaches should be prepared annually (Fig. 9.5).

9.3.7 Environmental Standards

Table 9.5 lists Iranian and European environmental standards for chemical, physical and microbiological parameters in surface water. There is evidence that these pollutants occur both in the vicinity of the Zayandeh Rud and in the river water itself.

The most important difference between Iranian and European Standards is the existence of yearly average values and yearly highest concentration in the European Standard. The European Directive also includes standards for parameters such as DDT, Benzene compounds and Hg. Standards for Cd depend on water hardness (European Community Directive 2008/105/EC). If concentrations of priority substances exceed the standards the water quality of the relevant stretch of river is classed as not better than moderate. Good ecological potential and good surface water chemical status according to the EU Water Frame Directive (2000/60/EC) has not been reached yet. Basically the good ecological potential should be reached for the entire Zayandeh Rud.

Table 9.5 Iranian and European environmental standards for priority substances in surface water

Parameter	Iranian Standard (mg/l)	European Standard yearly average (mg/l)	European Standard yearly highest concentration (mg/l)
Cyclodiene pesticides	–	$\Sigma = 0.01$	–
Cd[a]	0.1	$\leq$0.08–0.25	$\leq$0.45–1.5
Hg	–	0.05	0.07
Benzene	–	10	50
Ni	2	20	20
Pb	1	7.2	–
DDT	–	0.025	0.025
Mineral oil	10	–	–

[a]Cd–standard depends on water hardness

Iranian Standards for Benzene, DDT, and pesticides should be developed, as a priority.

Environmental Standards for COD (125 mg/l), BOD_5 (25 mg/l), total Phosphorus and total Nitrogen in wastewater discharged from wastewater treatment plants exist (European Community Directive 91/271/EEC 1991) and should be applied to Iranian wastewater treatment plants.

In terms of the European Waste Water Directive (Directive 91/271/EEC 1991) the Zayandeh Rud should be considered an extremely sensitive water body since the self-cleaning capacity of the river is limited. The fast flowing sections of the upper reaches have the highest capacity for self-purification. The downstream reaches of the river to the east have lower flows and reduced self-cleaning capacity while the organic pollution load increases. There is no dilution effect since tributaries are almost absent and the evaporation rate is very high, especially in summer. In addition water exchange with neighbouring water bodies (e.g. the sea) is not possible. In future the pollution load from wastewater should be limited to 2 mg/l total Phosphorus (10,000–100,000 PE) or 1 mg/l (>100,000 PE), and to 15 mg/l total Nitrogen (10,000–100,000 PE) or 10 mg/l (>100,000 PE).

Chemical water analyses of priority substances should be carried out regularly at each sampling station, consistent with laboratory capacities.

9.3.8 *Miscellaneous*

The oil company should be interviewed to determine whether the oil pipeline crossing the river upstream of Chamaseman Station is double walled or single walled: a double walled pipeline with a leak detection system is highly recommended to prevent possible oil spills into the Zayandeh Rud.

The introduction of water protection areas is recommended. According to EU Water Frame Directive (European Community 2000/60/EC), measures should focus on the protection against deterioration of water bodies that are used for abstraction of water for human consumption. Water abstraction at Chamaseman Dam is carried out from surface water, so safeguard zones should be established upstream. A specific hazard is use of roads near the river by tanker trucks carrying fuels and oils. Alternative routes away from the river should be considered. The use of water-endangering substances beside the river upstream of Chamaseman should be registered by the competent authorities, and control measures introduced for storage and use of these substances.

It is suggested that the two laboratories (Environmental Department and Water Board) should liaise with each other to ensure regular implementation of major chemical analyses, and that the chemical analyses of both laboratories should be compared and discussed in a common work group. The introduction of a quality management system is suggested to improve the quantitative and qualitative results of the chemical analyses, including inter-laboratory tests.

References

DIN 38410 (2004) German standard methods for the examination of water, waste water and sludge—Biological—ecological analysis of water (group M)—Part 1: Determination of the saprobic index in running waters

European Community (1991) Directive 91/271/EEC Council Directive concerning urban waste water treatment. http://eur-lex.europa.eu. Accessed 21 May 1991

European Community (2000) Directive 2000/60/EC of the European Parliament and of the Council of 23 October 2000 establishing a framework for Community action in the field of water policy. http://eur-lex.europa.eu. Accessed 23 Oct 2000

European Community (2008) Directive 2008/105/EC of the European Parliament and of the Council on environmental standards in the field of water policy. Accessed 16 Dec 2008

Hashemi H (2014) Calculation guideline of iran water resource quality index. Shahid Besheti University of Teheran, Environmental Research Institute

Hallock D (2002) A water quality index for ecology's stream monitoring program. http://www.ecy.wa.gov/biblio/0203052.html. Accessed Nov 2002

Länderarbeitsgemeinschaft Wasser (1998) Beurteilung der Wasserbeschaffenheit von Fließgewässern in der Bundesrepublik Deutschland

Mohajeri S, Nuñez von Voigt T (2011) Integrated water resource management in Iran. wwt-International, pp 29–32

Mohajeri S, Kaltofen M (2011) Bericht zum Tag der Hydrologie 2011 am 24./25. März an der TU Wien Fachgemeinschaft in der DWA. Hydrobrief: 53: 3

Varnosfaderany MN, Ebrahimi E, Mirghaffary N, Safyanian A (2010) Biological Assessment of the Zayandeh Rud River, Iran, using benthic macroinvertebrates. Limnologica 40:226–232

Chapter 10
Obstacles and Opportunities for Iran's Wastewater Sector Development

Shahrooz Mohajeri and Axel Dierich

10.1 Introduction

For 25 years the Iranian government has been seeking to improve and implement an integrated and consistent structure of the country's water and wastewater sector that is able to cope with its manifold disparities and challenges. These are, for instance, an increasing population density and urbanization rate, a rising standard of living and a continuous growth of the annual water consumption. Several concepts have been implemented and have led to a slight improvement of the sector as a whole. But there is still a high demand for optimization in order to improve the quality of services and to cope with future challenges. Regional Water and Wastewater Companies (WWC) face challenges particularly in the field of wastewater treatment.

In this chapter what factors prevent and which measures could help WWCs in shaping and implementing an economically, socially and ecologically viable wastewater disposal system will be discussed.

10.2 Procedure and Methods

The results presented in this chapter are based on a combination of several methods, which ensures comprehensiveness of the data collected. The data comprise our own expertise in the international and the Iranian wastewater sectors, a literature review, and 66 interviews with Iranian and international professionals from the fields of politics, administration and operation.

S. Mohajeri (✉) • A. Dierich
inter 3 GmbH Institute for Resource Management, Berlin, Germany
e-mail: mohajeri@inter3.de

S. Mohajeri, L. Horlemann (eds.), *Reviving the Dying Giant*,
DOI 10.1007/978-3-319-54922-4_10

Data on the main obstacles were evaluated and the results compiled with the help of a constellation analysis, which allows analysing and visualizing complex interactions of different factors (Schön et al. 2006), taking into account not only actors and institutions but also technologies, natural resources, and legal and economic factors. The results of the obstacles' analysis were empirically validated from the perspective of WWCs. Managers of the Iranian WWCs were surveyed in order to identify the obstacles of the Iranian wastewater sector from their perspective. The overall results were discussed and examined with Iranian and German professionals and supplemented with their ideas and suggestions. Based on this, some suggestions for greater autonomy of the WWCs were developed.

10.3 Initial Situation and Goals Set for the Iranian Wastewater Sector

Current efforts aiming at enhancing the Iranian wastewater disposal system are predominantly based on a new organizational design, as formulated by the Iranian government in the post-war year of 1991, and reflected e.g. in regular updates of the National Outlook Plan (Ardakanian 2005).

The most important, visible changes in the wastewater policy were the foundation of the Regional Water and Wastewater Companies (WWC) and the superordinate National Water and Wastewater Engineering Company (NWWEC). Until then, there had been no cohesive forms of wastewater disposal organizations (with the exception of a few individual towns). The scarce government resources that had been made available to the water and wastewater infrastructure during the war years had been invested mainly in the vital provision of drinking water. After the war, the wastewater sector suffered a considerable backlog in investment projects.

In this difficult situation, WWCs were supposed to be established in the form of modern organizations right from the start, featuring the following core elements (Manouchehri 2004, p. 57):

- Non-governmental status in order to abolish bureaucracy and foster motivation;
- Participation of the public in view of limited government resources and more effective implementation of plans;
- Decentralization by establishing self-governed companies and involving provincial municipalities to clearly separate governmental and administrative tasks;
- Consistent regulations for better coordination and control;
- Practical training for improving staff skills;
- Optimization as a result of the previous aims in the sense of increasing productivity and efficiency and providing better services at lower costs.

Furthermore, it was intended to

- Gradually meet demands by comprehensive development measures, and

- Implement a financial and economic system based on principles of industrial and cost accountancy in order to improve the financial position of the companies.

In the obstacles analysis, these and other measures were subjected to a critical review.

10.4 Obstacles for the Development of a Sustainable Wastewater Sector

What factors are keeping WWCs from shaping and implementing a modern and sustainable infrastructure of wastewater disposal? This was the key question of the obstacles analysis. The analysis was conducted predominately from the perspective of the WWCs. However, this perspective goes beyond the viewpoint of individual WWCs and comprises factors that impede modernization from an entrepreneurial perspective. The intention is not to present the full scope, but to focus on key obstacles and their interaction.

The obstacles that are hampering the development of the Iranian wastewater sector can be divided into four problem areas (see Fig. 10.1):

- Organizational deficiencies and law
- Management deficiencies
- Lack of financial means
- Water resources

These four problem areas will be discussed in greater detail below.

10.4.1 Organizational and Legal Deficiencies

With regard to organization and legislation, there are a number of obstacles that hamper efficient cooperation between the various institutions and companies of the

Fig. 10.1 Problem areas of the Iranian wastewater sector

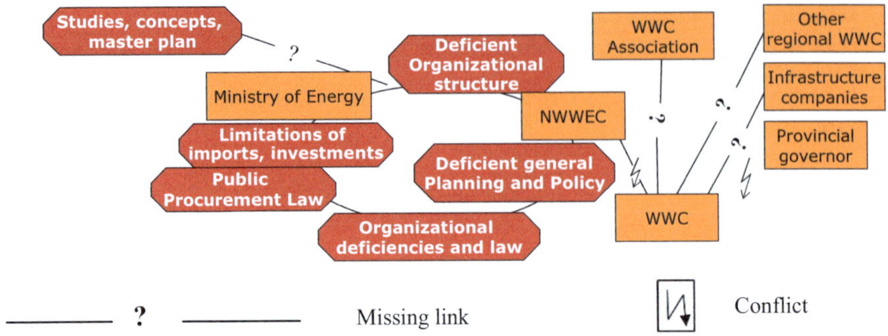

Fig. 10.2 Organizational and legal obstacles

wastewater sector and thus prevent the establishment of a professional system of wastewater disposal (see Fig. 10.2).

The new organizational structure introduced in 1991 has been implemented only in part, which is why the positive effects that were hoped for did not meet expectations (UN 2004). WWCs have proved their value as new institutions, with increasing technical knowledge and working capacity at provincial level and increasing independence from central government. Yet the overall organizational structure has remained centralistic, with features of a planned economy that lead to inefficiencies and misallocations in the entire wastewater sector (The World Bank 2007).

Cooperation neither between the NWWEC and the WWC (vertical) nor between the provincial governments and WWCs (horizontal) is ideal. According to the results of the analysis, intentions of the NWWEC are not systematically communicated to WWCs, and communication of their suggestions to NWWEC, which could contribute to solving problems and optimizing the wastewater sector, hardly takes place. At the regional level it was found that cooperation between the WWCs and the provincial governments was underdeveloped. Urban WWCs do not sufficiently cooperate with rural WWCs, wastewater infrastructure and urban planning are not coordinated, and apart from this, there are neither incentives nor possibilities of coordinated cooperation between the different local and regional infrastructure operators. The poor cooperation between WWCs and the provincial government, for instance, becomes obvious in delayed building permits.

Administrative separation of water resources and water infrastructure within the Ministry of Energy have prevented the formulation and implementation of a consistent national water and wastewater policy including a commitment to goals or duties and rights defined for WWCs and other institutions involved in the vertical and horizontal organizational structure (The World Bank 2007; Scholz 1992).

Universities focus on theoretical research and even the Office for Training, Research and Technology of the Ministry of Energy cannot adequately provide for requisite knowledge at the operational level. There is high demand but little financial means for know-how, also from foreign countries. It is, however, still difficult for foreign consultants and companies to compensate for these knowledge deficits. Although foreign investment has been eased by a reduction of bureaucratic hurdles and the strengthening of investment protection, there is still no remarkable international involvement in Iran's wastewater sector.

Moreover, legal framework conditions limit the availability and use of technologies and products. The contract award regulations, for instance, stipulate that offers are assessed based on 80% economic criteria and only 20% technical criteria, and usually the most economical tender is awarded the contract. WWCs are therefore nearly forced to implement projects featuring poor technical quality.

Import restrictions that were actually meant to protect national production had adverse side-effects and too few products of excellent quality were offered on the domestic market.

One of the basic obstacles in the area of organization and legislation is the lack of a reliable, coherent and consistent wastewater policy provided by the national government. As long as there is no financial or technical master plan for the development of the wastewater sector, WWCs remain at the mercy of individual, case-related decisions of the NWWEC which are not always transparent or comprehensible, as the survey suggests.

WWCs lack the basis for medium-term business strategies featuring viable financial and investment planning. The national priority list for the advancement of wastewater infrastructure projects compiled by NWWEC has not been implemented yet. The imperative of equal distribution of financial means between regions outweighs water management goals.

The lack of political and administrative accountability, the deficiency of knowledge – insufficiently used and accumulated in the wrong places – as well as counterproductive regulations, constitute serious obstacles to the modernization of the wastewater disposal system. This causes numerous follow-up problems of service quality and financing and it has a demotivating effect on WWC management and staff.

10.4.2 Management Deficiencies

The effects of the unfavourable organizational and legal structure of the Iranian wastewater sector become evident and tangible in the area of operative management (see Fig. 10.3).

The deficits in operative management derive from the overall lack of planning in the wastewater sector at the national level, the largely curtailed scope of action of WWCs, and the lack of integration in financial, technical and know-how networks. As a result, WWCs are not capable of developing their own local or regional wastewater disposal concepts as a strategic basis for their business activities. Under these conditions, wastewater disposal remains piecemeal and fails to fulfil its potential.

There are, however, also management deficits that are accredited to the WWCs themselves. In the past, a serious obstacle was poor planning of plants and

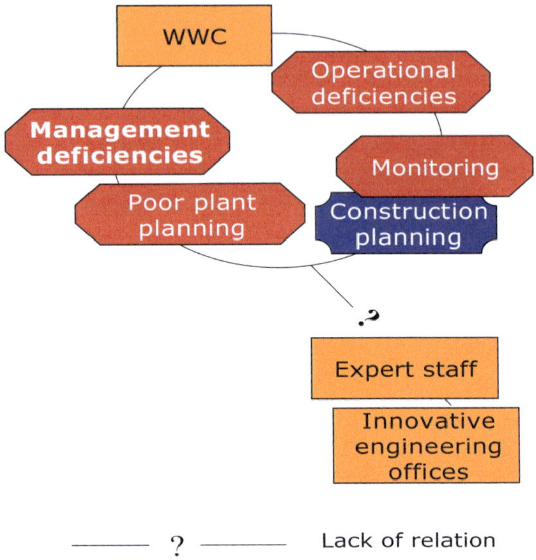

Fig. 10.3 Obstacles in the area of operative management

insufficient supervision of building projects. Thus, despite a shortage of funds, wastewater treatment plants are often over-dimensioned or sewages were built without connecting users. Well-planned plants are frequently operated uneconomically due to operational deficits such as poor process optimization or faulty technologies. A lack of qualified personnel is an important obstacle for the development of the Iranian wastewater sector (see previous section). This situation gets even worse as the varying hydrogeological conditions in different Iranian regions require specific regional problem-solving instead of standard solutions of wastewater disposal (Otterpohl 2008). Yet it seems that there are too few innovative engineering consultancies that have the relevant know-how and proactive attitude to provide such solutions.

The management of WWCs can neither build on reliable overall planning by the Ministry and the NWWEC nor on qualified workers or solution-oriented engineering consultants, and therefore cannot develop or implement viable strategies concerning wastewater management or technologies.

10.4.3 Lack of Financing

The unreliable and insufficient financial basis is a serious obstacle to the modernization of the Iranian wastewater sector. This became evident from the interviews with WWC managers who –unsurprisingly– gave top priority to the problem of finance.

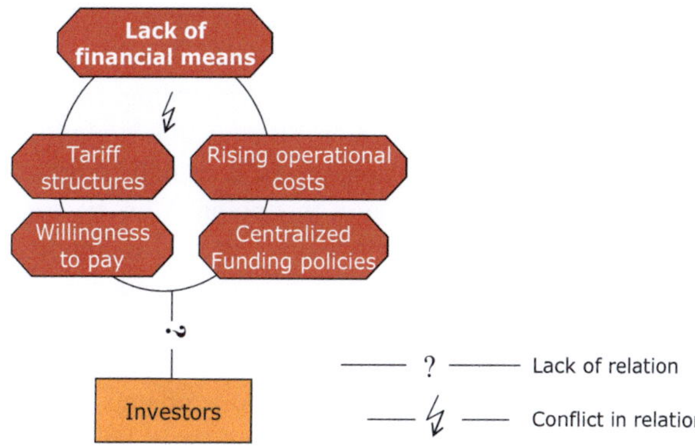

Fig. 10.4 Obstacles in the area of financing

In this problem area, several factors come together that inhibit sound budgeting and investment planning and reliable operations (see Fig. 10.4).

The structure of tariffs does not reflect the accruing costs of wastewater disposal and indeed tariffs do not cover costs. Hence, fees do not help to refinance the fixed costs but are usually used only for covering running expenses. On the one hand, a few years ago, the initially acquired authority of the city councils was undermined by the central government's interference with the politically motivated shaping of charges and tariffs. On the other hand, fees cannot cover the costs as long as WWCs create increasing operating costs themselves as a result of inefficient management, including in particular inefficient use of materials and under-utilization of the plants. In the survey and interviews, managers criticized inefficiency and tariffs as not being cost-covering, yet they did not link the two together. Inflation further exacerbates the problem of operating costs.

Unsatisfactory and poor economic management leads to companies not having enough of their own funds to invest in network and plant construction but depend on government grants and subsidies. In fact, subsidies for plant construction are common by international standards, but modern infrastructure policies only provide for paying these investment grants based on a company's long-term local or regional operation and financing concept. Concepts of this kind are not demanded of companies in Iran. Even worse: on request of the WWCs, the providers of subsidies, i.e. the Ministry of Energy and the provincial government, hand out project-related funds at their own discretion, in a centralized and planned economy type procedure. Due to the tight budgetary situation, these project-related investment grants are stretched over several fiscal years so as to keep the annual burden on the budget within limits. This inefficient and unreliable financing situation leads to extremely long construction periods, which in turn increases costs and reduces the quality of building projects. This financing procedure provides no incentives for efficient use of funds in the regions.

Tapping additional, non-governmental funds, which is an integral part of financing in modern infrastructures, is not common in Iran. Domestic investors consider the wastewater disposal business to be unprofitable and foreign investors hesitate because they perceive the investment climate as being risky while their potential investments are hampered by legal restrictions.

The fourth pillar of infrastructure financing, i.e. loan financing, cannot be fully exhausted either. Both the World Bank and Islamic Bank grant loans to WWCs, but not the Iranian banks. Whether this is an indication of the WWCs' lack of creditworthiness remains unclear at this point. After all, granting loans to infrastructure operators (especially local ones) is certainly common in other countries.

WWCs essentially have four potential sources of income: fees, investment grants from the Ministry of Energy and provincial governments, and loans. While fees are only a small, albeit calculable source of income, subsidies granted by the Ministry of Energy and provincial governments are defined from year to year and cannot be used as a basis for medium-term, let alone long-term investment planning. The companies therefore have an extremely restricted financial scope of action, which in addition is irregular and unpredictable. As a result, they can neither establish nor implement medium-term or long-term wastewater management strategies and have no incentive to improve the unprofitable wastewater disposal situation at their own regional or economic initiative. The fact that private participation or private operator models have not been implemented yet could be the result of a lack of investor confidence in the sector in general or one of the reasons for inefficient management.

10.4.4 Attitude to Water Resources

Until recently, an additional obstacle was the lack of awareness even among political actors, planners and engineers about water resources scarcity and the potential for wastewater reuse. The consequences of this lack of awareness became evident especially in a conflict between the Ministry of Agriculture and the Ministry of Energy about the use of treated wastewater for irrigation of agricultural acreages. About 10 years ago, the poor quality of wastewater treatment had caused the Ministry of Agriculture to issue a ban on its use for agricultural irrigation, which in turn contributed to the overstraining of freshwater resources.

This has however changed due to the paradigm shift that took place in 2013/2014 in political handling and public communication of the severe and aggravating water scarcity problems. Interestingly, wastewater has even become a high valued resource and conflicts now revolve around the question of the legitimate owner of (treated) wastewater. However, in order to become a significant alternative resource, the treatment of wastewater needs to be improved on a large scale.

Further possibilities for increasing resource efficiency, such as using wastewater as fertilizer or for biogas production, are at the most considered in theory, but far from being applied in practice. Here, too, limited cross-sector cooperation and concepts constitute an obstacle as illustrated in Fig. 10.5 with the Ministry of Agriculture as one example.

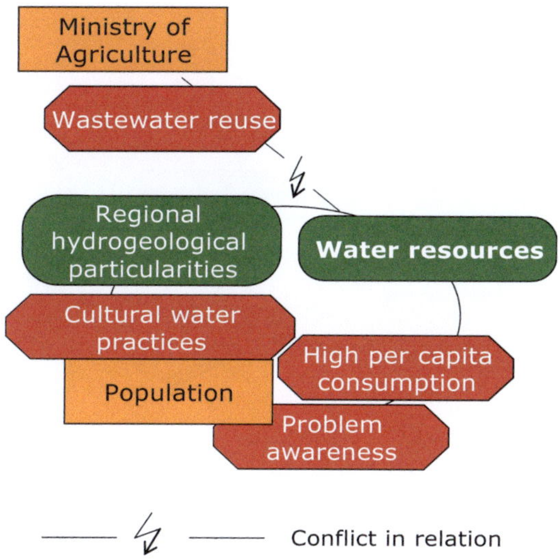

Fig. 10.5 Obstacles in the area of water resources

While awareness for wastewater as a resource is developing in the minds of professionals and decision-makers, the population still lacks this awareness which is reflected in comparatively high per-capita consumption. With a growing population, water consumption and wastewater generation are steadily increasing (Tajrishy and Abrishamchi 2005). At the same time, awareness of proper wastewater disposal is not very distinct. Consequently, the people's willingness to pay for wastewater disposal is comparatively low.

The different regional hydrogeological conditions also constitute an obstacle to wastewater disposal infrastructure development in Iran. On the one hand, less expensive and standardized disposal systems are not appropriate, while on the other hand, companies and regions lack political and financial autonomy and governmental investment is still too low for developing and implementing more complicated but regionally adjusted disposal concepts. Already scarce funds are often invested in structures and technologies that lag far behind current possibilities of cross-sector resource protection and resource efficiency.

In view of the currently extreme lack of water in many parts of the country, the involvement of the public sector in the systematic treatment and reuse of wastewaters is starting to become a key element of Iran's water policy. So called buyback contracts are becoming increasingly common. Industrial or agricultural businesses, but also individual people, build wastewater treatment plants with their own funds and in return get a 20-year right of disposal of the treated wastewater.

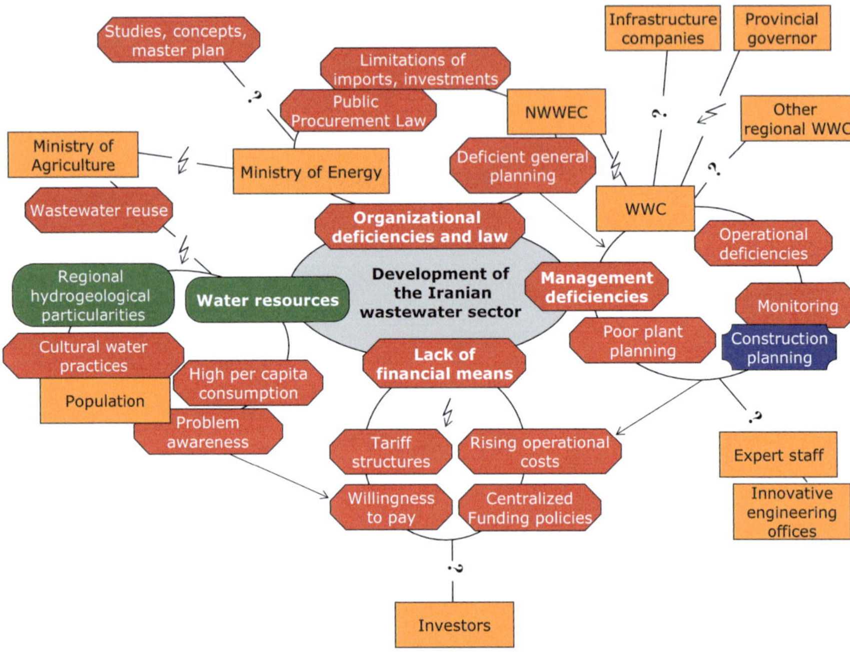

Fig. 10.6 Overview of all obstacles

10.5 Summary of Obstacles

Each of the four problem areas is made up of several individual elements and is interlinked in several ways (see Fig. 10.6). This must also be kept in mind when interpreting the results of the survey: the majority of Iranian WWC managers see the greatest need for action in the elimination of financial hurdles, followed by the elimination of organizational, legal and operative deficits. However, this ranking must not hide the fact that all of the areas need to be tackled similarly.

A strategy for overcoming these obstacles must take into account that the various elements interact and must consequently tackle all of these areas. It is of no use if financing, for instance, is improved, yet the funds made available are used inefficiently because of organizational, planning or operational deficits. Any new course of a financing strategy must be accompanied and supported by the respective organizational and legal reforms, and ultimately also be effectively implemented at the operational level. Moreover, convincing the population of the necessity of systematic wastewater disposal and making them agree to pay disposal fees can only work if the modernization of local wastewater disposal systems appears as a targeted, efficient and successful undertaking. Hence, the different levels – national and provincial government, WWCs and the population – need to be included in a consistent overall strategy and their respective needs and scopes for action taken into account. A holistic and integrated organization of the water and wastewater sector is strategically important to lead resource policies to success (Tejada-Guibert and Čedo 2001).

The reform and modernization measures decided by the Iranian government in the 1990s (see Sect. 10.3) all pointed in the right direction but have still not been implemented to a sufficient degree. The fact that Iran is in principle on the right path is also reflected in its commitment to Integrated Water Resources Management (IWRM). To support and speed up the development of a modern and viable wastewater infrastructure, the obstacles analysed here, however, need to be eliminated with determination and dedication.

Important approaches for overcoming these obstacles are

- Targeted structural linking of various action levels and political spheres (horizontal and vertical integration of institutions and actors) based on an integrated water policy (Young 2002),
- Communicating an integrated water policy to the professional public (rules and regulations, know-how) and providing targeted information and involving the broad public (building acceptance, using local knowledge) (Newig 2005),
- Rearranging the strategic, financial and operative scope for WWCs' actions aimed at implementing an integrated water policy.

10.6 Reorganization of Decision Making, Governance and Responsibilities: Towards the Wastewater Sector of the Future

Propositions for the reorganization of the water and wastewater sector in this section are made against the particular historical background. They try to catch up with what was originally conceptualized in the reform process in terms of decentralization, autonomy and private sector participation. On the other hand they take into account the fact that merely delegating responsibilities to the local level and private sector cannot succeed without supervisory mechanisms at the national level.

10.6.1 Decentralization

One of the main objectives of decentralization is the shifting of tasks to the operational level of those involved in it (see Edwards et al. 1993). This objective is closely connected to the creation of greater autonomy for WWCs. The state – instead of becoming involved in the actions of WWCs – should concentrate on core tasks in the water sector like

- Setting political guidelines and general objectives,
- Monitoring the realization of guidelines and objectives,
- Creating incentives to support the implementation of political guidelines and general objectives.

10.6.1.1 Political Guidelines and Overall Objectives

The state should set a more definite course with regard to political guidelines, specifying how the Iranian water and wastewater sector is to be developed. Important objectives include equalizing living conditions and generating sustainable resource protection. The task of the state is to communicate political objectives to WWCs and to develop the necessary framework and instruments for their implementation.

The state needs to set clear standard parameters to achieve these objectives. This applies mainly to drinking water quality, hygiene, resource/environmental protection and price setting. Instruments include institutions for control, surveillance and punishment of noncompliance, as well as financial incentives. Meeting standards should be, however, the responsibility of WWCs, not the state, irrespective of the nature of the ownership and the size of the service area.

10.6.1.2 Direct Control: Autonomous Control Bodies

Privatization of the water and wastewater sector needs to be accompanied by an autonomous control regime: In the UK and Wales the water industry was privatized in 1989 and in Chile in 1998. Their initial situation was similar to the situation in Iran we witness today. Arguments in favour of privatization included increased competition, greater efficiency of the private sector and greater investment ability of private companies for repairing water systems and to meet new water quality standards. As a precondition for success unique control regimes were set up in these countries.

Based on the conviction that water provision and wastewater disposal are publicly provided goods and thus a responsibility of the state, it is recommended that in Iran a well-functioning network of state control mechanisms should be implemented.

10.6.1.3 Indirect Control: Financial Incentives for WWC

An array of further instruments is available to the state for supporting designated political guidelines and objectives. Chief among these is the deployment of financial means for creating incentives for WWCs. Conversely, the state can also introduce a negative incentive system, e.g. by charging fees or fines.

State subsidies are used internationally as a positive incentive system for motivating companies to quickly implement political objectives. The decision for allocating financial funds should be tied to the performance of WWCs, to generate competition and consequently increase efficiency and customer satisfaction. Using suitable benchmark systems, the state can create transparent systems for allocating funds while effectively increasing the quality and transparency of WWCs. To this end, transparent and homogenous criteria can be used to check how WWCs perform by national or international comparison.

International experience, like the introduction of monetary incentives in Germany's new Länder (states) after the reunification, shows that taxes and fees can be used to achieve political objectives such as improved water consumption management or water pollution protection (König and Heimann 1998). For instance, the use of suitable fees for surface and groundwater extraction could contribute to lowering water consumption to the extent that charges are passed on to customers. If revenues from those fees are used for environmental education or for financing water-saving faucets, efficient water consumption management could be developed.

10.6.2 Increasing Autonomy

A key theme of recent reforms in public management in various countries and sectors is the perceived need for many organizations to have a degree of legal, financial and operational autonomy. In organizational theory, autonomy is generally seen as right permitting an organization or person to take and implement decisions on their own without imposed obedience to external actors taking part in those decisions (Horn 2002). Accordingly, autonomy of WWCs is seen as one of the determining variables for an effective water supply and sewage disposal.

With more autonomy and responsibility transferred to WWCs, a clear separation of tasks and responsibilities should be realized between the state and regional WWCs. These need to be put in a position to independently make necessary decisions for efficient and customer-oriented operation and to develop strategic company planning. Hence, while the state needs to control WWCs efficiently through developing a sustainable water management policy, WWCs need to be free to act and be fully accountable within their areas of responsibility.

10.6.2.1 Legal Autonomy

Legal autonomy implies that independency is formally granted by law. This is almost the case for urban WWCs today; however, in detail the legal framework for the water and wastewater sector as a whole enables the national government to exert political influence in various ways.

For example, WWCs do not have sufficient authority to decide and pass their own rules. These are provided by the ministry which does not take into account local conditions. WWCs must therefore be enabled to write and ratify their own charters within the framework of the national legal system, stipulating e.g. the legal form, the service area, the procedure for electing and forming a general assembly and an executive board, their obligations and range of activities or agreements concerning capital stock. Furthermore, the charter must define the tasks of the WWC which are dedicated to the following objectives:

- Efficient, reliable and faultless distribution of drinking water in hygienically acceptable condition to the connected population in the respective area of provision,
- Efficient, reliable and faultless disposal of wastewater for all customers connected to sewerage, with the aim of extending the connection rate to the greatest possible extent,
- Economically maintainable, reliable, trouble-free treatment of all collected wastewater to a quality level that guarantees the preservation of the natural ecosystem of the receiving waters over the long term,
- Long-term and sustainable planning of all tasks and measures and the coordination with other relevant actors within and outside the area of provision.

Apart from this, general assemblies and executive boards of WWCs should be established that have the sufficient legal scope to independently decide on the organizational structure of WWCs and on questions of restructuring. In doing so, WWCs could take into account specific local requirements such as the number of customers, the size of the provided area, climatic conditions and local socio-cultural characteristics. Decisions on the structure and design of the different divisions, their merging or the creation of new divisions need to be taken exclusively by WWCs. Decisions on the kind and extent of private participation in the company should also be taken by a WWC's executive board and general assembly.

Moreover, the WWC managers interviewed criticised their lack of authority to sanction illegal water usage and that clear enforcement rules are inexistent. On the contrary, they said the lack of regulations has led, in some provinces, to a promotion of illegal usage instead of providing accurate solutions for managing the relationship between WWCs and customers.

10.6.2.2 Policy Autonomy

Policy autonomy refers to the ability to autonomously pursue strategies with regard to internal as well as external company issues, to represent these strategies externally and to implement them. This applies mainly to company planning as well as the merging of interest representation, which is performed against the backdrop of the requirements of the company and its customers. These strategies and actions must, of course, be in accord with the current legal framework as well as the water and wastewater sector objectives and guidelines as defined by the state. Policy autonomy is thus manifested on two levels: internal autonomy and external autonomy.

In the first place, WWCs need to be free to develop strategic business plans and to decide on the company structure and target setting regarding privatization activities, human resources, optimization (of consumption, purchase, monitoring, repair etc.) or the scope of business activities. It must be ensured that the WWCs' actions make a clear contribution to the realization of state-planned objectives while also bringing about a high level of customer satisfaction.

For a modern water and wastewater sector in Iran, a solid network of WWCs is necessary. This should be facilitated and financially supported by the state. The creation of a WWC association for representing their interests could enable them to act robustly and in unison in relation to business partners, officials and governmental authorities. Water management expert associations could allow WWCs to systematically treat and develop solution concepts for technical issues in the water and wastewater sector in cooperation with the private sector and the scientific community. In Iran, there exist a couple of expert associations but there is no association that bundles the interests of WWCs. This inhibits WWCs from articulating their interests at the political level.

10.6.2.3 Financial Autonomy

The basis of business is the freedom to set prices for the product or service offered. This only applies to a limited degree to services such as water provision or wastewater disposal as these services constitute a monopolistic public good. For this reason, prices should be set in close and constructive cooperation between WWCs and public bodies.

Step by step WWCs should be put into a position to set their prices so that revenues provide a secure basis for operation and re-investment in infrastructure. WWCs must also be able to decide on how to use their revenues. This is especially important in view of the efforts toward a privatization of WWCs and should be consistently pursued and implemented.

Along with this, the state should ensure professional regulation (The World Bank 2007). The government should develop and establish adequate controlling institutions and methods to analyse, compare and evaluate the prices proposed by the WWCs. Only after prices are approved by a controlling institution can the tariffs come into effect.

WWCs need price security; whether revenues are gained directly from the customers or from a regular subsidy by the state is a political decision. By contrast, WWCs must be able to ensure and prove that revenues from fees or from subsidies are used exclusively for the purposes intended. For instance, revenues from the sale of connections or fee revenues for re-investment should be used exclusively for investment purposes and not for covering operational costs.

Financial autonomy, however, can only realize its potential if necessary basic conditions are found in the company. One of the most important basic conditions is that the company is free of debt. This is currently not the case for Iranian WWCs. In order to achieve a debt-free state, the following is proposed:

- Transfer of the ownership of water and wastewater facilities, equipment and other property to WWCs. By transferring the ownership from NWWEC, municipalities or regional water organizations to WWCs, some amount of their debt could be reduced. To do that, an emergency law is necessary, which liberates WWCs from paying taxes on the transfer of ownership.
- Reset property assessment. Non-public urban water and wastewater companies are usually not allowed to evaluate their properties without tax. The property tax that will be accrued is 50 to 60%. Recently, this agenda led to increased capital and reduced the cumulative loss.
- Deferment of debts. According to existing law, the state can defer 50% of the debts of urban WWCs, which were incurred as a result of state subsidies for investment in water and wastewater facilities. The other 50% should be able to be paid back as an interest free loan over a time period of 10–30 years.

Such measures would by and large allow WWCs to establish financial autonomy and also encourage the interest of private partners for the company.

It is to be assumed that WWCs will not be in a position to handle large investments on their own in the near future. Over the medium term, however, they will be able to assume responsibility for investment by drawing on revenues (re-investment). Until the desired situation is reached, the state would have to make necessary investments.

10.6.2.4 Managerial Autonomy

Managerial autonomy refers to the ability of company management to autonomously make decisions and act within the existing legal framework and in the interests of an efficient and sustainable water provision and wastewater disposal.

With this in mind, a regionally appointed general assembly should appoint an executive board of the WWCs in the future. The executive board should be responsible for the operation of the company. Depending on the charter, the board could make decisions independently or in unison with the general assembly on financial, technical and strategic issues for water provision and wastewater disposal as well as for the development of the company.

In order to cope with responsibilities transferred to the WWC management, far-reaching freedom to act is indispensable for company management for decisions concerning human resources (development). Decisions regarding the adequate training requirements and the selection of personnel should be taken by the WWC management (see Chap. 5 CD). This would ensure that the company has enough well-trained personnel prepared for future challenges. In order to obtain enough well-trained personnel for the company amidst the competition in the labour market, it is necessary to grant management more freedom in negotiating salaries.

10.6.3 Private Sector Participation (PSP)

From a governmental perspective the most complete forms of decentralization are privatization and deregulation (also called "market decentralization") because they shift a considerable degree of responsibility from the public to the private sector. They allow functions and decisions that had been primarily or exclusively the responsibility of the government to be carried out by private or semi-private businesses, with or without the involvement of investors, banks, private associations and other non-governmental financial donors.

For the Iranian state the transfer of tasks to private actors means that it withdraws from the responsibility of fulfilling certain tasks itself. This of course implies that it loses influence on many decisions on operative matters. Yet the state does not abdicate complete responsibility for the water and wastewater sector. State authorities (local, regional or national – depending on the degree of decentralization) retain influence over strategic and framework decisions and must put even more emphasis on the control of target achievements by water supply and wastewater disposal.

The lack of governmental financial means and the necessity of developing water and wastewater infrastructure in Iran have been the main reasons for the government to reduce investments and foster PSP. Therefore, a lot of opportunities for PSP have been created during the last years. Since the withdrawal of the World Bank from Iran, the Islamic Development Bank (IDB) has become one of the most important investors for the Iranian water sector, with total loans of around 862 m Euros.

Moreover, international and domestic loans and investments have been promoted, e.g. foreign (particularly Chinese) investments of around 900 m and domestic investments of around 185 m Euros. Between 2011 and 2013, more than 220 m Euros were invested in water and wastewater infrastructure by the Iranian government in the form of government bonds.

Private sector partnerships in the form of BOT (build-operate-transfer), BOO (build-own-operate) or buyback contracts are becoming increasingly important. Within a few years, 380 m Euros were generated and invested by the private sector for water and wastewater infrastructure projects, 330 m for desalination projects and 660 m for buyback projects.

Increasing PSP requires the state to set clear objectives, guidelines and standards obligatory for all WWCs regardless of their organizational form or ownership status. After the PSP has started, mechanisms of evaluation, control and sanctioning need to kick in. While control should be in the hands of autonomous control units, the government remains responsible for evaluating the target achievement. To this end, it benchmarks the WWC and the PSP, also in order to create transparency and public awareness for their performance. WWCs themselves are required to cooperate in benchmarking, act in a transparent and accountable way and commit themselves to efficiency by keeping costs and tariffs as low as possible while guaranteeing high quality levels.

References

Ardakanian R (2005) Overview of water management in Iran, In: National Research Council (eds) Water conservation, reuse and recycling. Proceedings of an Iranian: American Workshop, Washington DC

Edwards DB, Rosensweig F, Salt E (1993) Designing and implementing decentralization programs in the water and sanitation sector. WASH Technical Report No. 89, USAID, Washington DC

Horn MJ (2002) The political economy of public administration: institutional choice in the public sector. Cambridge University Press, Cambridge

König K, Heimann J (1998) Sieg der Üblichkeit: Wasserversorgung und Abwasserentsorgung. In: Czada R, Lehmbruch G (eds) Transformationspfade in Ostdeutschland. Beiträge zur sektoralen Vereinigungspolitik, Frankfurt a. M./New York

Manouchehri G (2004) Die Entwicklung des iranischen Wasserversorgungs- und Abwasserentworgungssektors–Vergangenheit. Gegenwart und Zukunft, Aachen

Newig J (2005) Die Öffentlichkeitsbeteiligung nach der EG-Wasserrahmenrichtlinie. Hintergründe, Anforderungen und die Umsetzung in Deutschland. In: Dierkes M et al (eds) Zeitschrift für Umweltpolitik und Umweltrecht, Frankfurt am Main

Otterpohl R (2008) Cost-Efficiency in water management through demand side management and integrated planning. In: Baz A, Ismail et al (eds) Efficient management of wastewater. Its treatment and reuse in water scarce countries, p. 205, Germany

Scholz O (1992) Wasser und Abwasser in einer Hand – eine unternehmenspolitische Entscheidung. In: GWF Wasser/Abwasser Jg. 133 (1992) Nr. 11, Munich

Schön S, Kruse S, Meister M, Nölting B, Ohlhorst D (2006) Handbuch Konstellationsanalyse. Ein interdisziplinäres Brückenkonzept für die Nachhaltigkeits-, Technik- und Innovationsforschung, München

Tajrishy M, Abrishamchi A (2005) Integrated approach to water and wastewater management for Tehran, Iran, In: National Research Council (eds) Water Conservation, Reuse, and Recycling. Proceedings of an Iranian-American Workshop, Washington DC

Tejada-Guibert AJ, Čedo M (eds) (2001) Frontiers in urban water management: Deadlock or hope? IHP-V Technical Documents in Hydrology No. 45. UNESCO, Paris

The World Bank (2007) Making the most of scarcity: accountability for better water management results in the Middle East and North Africa.: http://web.worldbank.org/WBSITE/EXTERNAL/COUNTRIES/MENAEXT/0.contentMDK:21244687~pagePK:146736~piPK:146830~theSitePK:256299.00.html Accessed July 2016

UN (2004) Freshwater and sanitation country profile. Department of Economic and Social Affairs, Division for Sustainable Development, The Islamic Republic of Iran. http://www.un.org/esa/agenda21/natlinfo/countr/iran/Iranwatersanitf.pdf Accessed July 2016

Young O (2002) The institutional dimensions of environmental change. fit, interplay, and scale. Cambridge University Press, Cambridge

Chapter 11
Vulnerability Analysis of Farmers in the Roodasht Region, Iran

Wolf Raber, Parisa Jafari Berenji, Mohammad Naser Reyhani, Shahrooz Mohajeri, and Lena Horlemann

11.1 Introduction

Climate change and water resource overuse have led to water scarcity in Iran over the years. This puts high pressure not only on Iran's environment but on sectors with high water consumption, like agriculture in particular. Agriculture does not only overstrain surface- and groundwater resources by excessive water withdrawal for irrigation but also by heavy pollution of the Zayandeh Rud river with agricultural inputs.

The area under investigation is located in the lower reaches of the river (see Fig. 11.1) and comprises the populated and cultivated lands served by the Roodasht irrigation network and nearby wells.[1] Roodasht borders on the Dasht-e Kavir desert in the north and east, and the Gavkhuni, a salt lake and marshland designated as UN Ramsar site, in the south-east. The population of the area is about 50,000 including more than 10,000 active farmers and their families.

After the Iranian revolution modern irrigation systems (Abshar and Roodasht network) were built to increase agricultural production, and in years with sufficient water resources the canals provide water for around 70,000 ha of cultivated area.

Climate change and resource depletion have resulted in extreme water shortage and ongoing desertification in the region. Ironically, cultivation has prevented the expansion of the desert at the same time. Drought periods, however, have brought the government to rationalize and frequently cut surface water supply to Roodasht.

[1] Although it is not its official (administrative) name of this region, for more convenient reading we will call it just "Roodasht" in the text but it is actually the "region served by the Roodasht irrigation network".

W. Raber (✉) • P. Jafari Berenji • M.N. Reyhani • S. Mohajeri • L. Horlemann
inter 3 GmbH Institute for Resource Management, Berlin, Germany
e-mail: raber@inter3.de

S. Mohajeri, L. Horlemann (eds.), *Reviving the Dying Giant*,
DOI 10.1007/978-3-319-54922-4_11

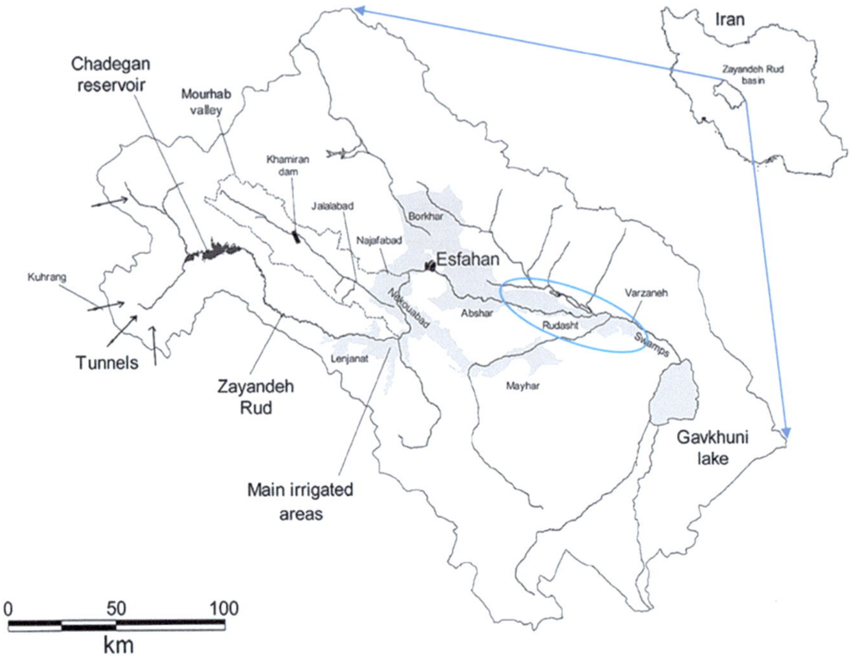

Fig. 11.1 Location of the Zayandeh-Rud basin and Roodasht region (source Molle et al. 2009)

As a result, farmers started to overuse local groundwater resources, which again led to dropping water tables and salinization.

Tensions between different water users have been increasing. Though farmers have received some financial compensation for their crop losses, all these measures are in no way sustainable or able to substitute a long-term strategy for land and water management. In order to develop a feasible and acceptable land use concept that helps stakeholders prevent or reverse land degradation and use water resources efficiently, it is necessary to depict the complex situation. A vulnerability assessment is an adequate tool to understand the sensitivity of farmers to water scarcity and possible starting point for developing adaptation measures.

11.2 Theoretical Framework and Method

Vulnerability is often conceptualized as the degree to which a system is susceptible to harm when exposed to hazards and risks (Turner et al. 2003; Adger 2006; Füssel 2007). However, in most cases, vulnerability itself is a consequence, resulting from underlying social and environmental conditions which need to be understood (Voss 2008).

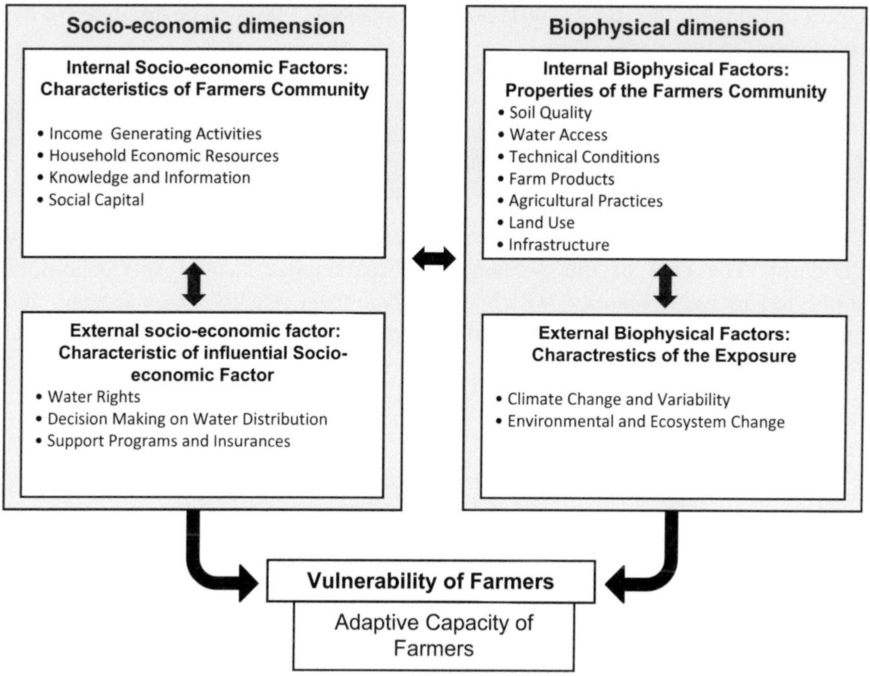

Fig. 11.2 Applied vulnerability framework (adapted from Füssel 2007)

Against this backdrop, the theoretical framework used to analyse the vulnerability of farmers to water scarcity in Roodasht is based on a comprehensive approach presented by H.M. Füssel (2007). According to Füssel, vulnerability is conceptualized by distinguishing among the internal and external spheres (or scales) and the socio-economic and biophysical knowledge domains. The internal sphere includes factors describing the properties of the vulnerable system or community itself, whereas factors from the external sphere describe aspects beyond the local community. By adapting Füssel's classification scheme, 16 factors to assess vulnerability of farmers to water scarcity in Roodasht have been classified and assorted to the overall structure of the vulnerability framework Fig. 11.2.

In view of these 16 factors, data collection and analysis was facilitated. It was based on a mixed method using both qualitative and quantitative data which include more than 20 site visits, a literature review, analysis of official data, 30 in-depth interviews with farmers, seven workshops and 30 interviews with experts from administration and civil society, and a remote sensing analysis.

Qualitative data of farmers' interviews and meetings with the experts were analysed topic by topic. Regarding selected internal and external factors, the data was subsequently examined, coded and categorized to identify relevant topics and issues. Extensive field notes collected during the site visits were used throughout the analysis and coding. Based on this, the main thematic fields of vulnerability in Roodasht were identified (see Sect. 11.3.1). Adaptive capacity as an important aspect of vulnerability (IPCC 2001; Smit and Wandel 2006) will be discussed in the conclusions only.

11.3 Vulnerability Assessment

This section presents the current state of vulnerability of farmers to water scarcity in Roodasht. The goal of this section is to characterize factors and conditions connected to water scarcity, which currently impact negatively on farmers and may ultimately lead to environmental degradation, desertification and social conflicts.

11.3.1 Thematic Fields of Vulnerability

The main cause for vulnerability of farmers in the region is their exposure to poor *water availability in Roodasht* during the last water scarce decade. Coupled with a range of connected factors, the lack of irrigation water in years gone by has led to significant cuts in local *agricultural production*, impacting negatively on *farmers' livelihoods*. Amongst others, the aspects mentioned have caused *regional entrepreneurship* to stagnate and restrict activities on *environmental conservation*. Adversely, the absence of ecologic and sustainable regional development negatively affects on-farm activities and the livelihood of farmers.

It has been found that in the past decade farmers have become increasingly vulnerable to water scarcity in Roodasht. With no structured and powerful counteractive measures, the main risks of *desertification and environmental degradation* as well as *social conflicts* are expected to lead to a crisis in the near future, while symptoms can already be observed today.

In Fig. 11.3 the thematic fields with associated conditions are found in the green boxes and characteristics in Roodasht in the blue boxes. Due to limited space in this chapter, only the main thematic fields will be discussed with selected aspects in detail.

11.3.2 Water Availability in Roodasht

In the past decade, the effects of climate change and particularly climate variability have led to variance in precipitation[2] and higher temperatures in the Zayandeh Rud basin and specifically in Roodasht; a phenomenon which is expected to continue in

[2]Natural precipitation in Roodasht is usually lower than 100 mm/a.

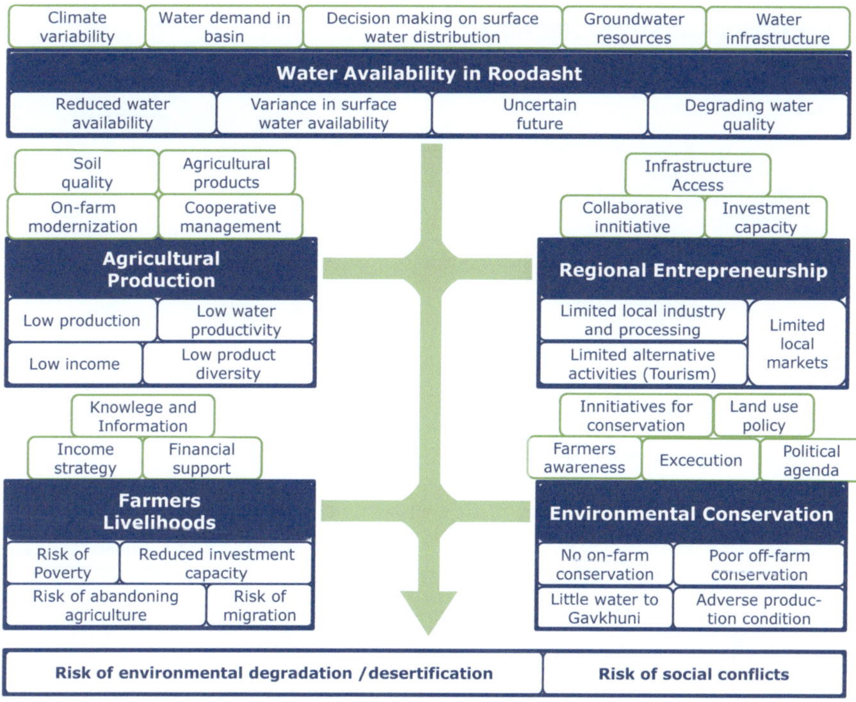

Fig. 11.3 Thematic fields of farmers' vulnerability in Roodasht

the future (Gohari et al. 2013; Eslamian et al. 2016). At the same time, the number of water users in the basin has increased due to population growth, economic development and intensified agricultural activities. The reduced water availability and increased competition for scarce water resources has led to significant cuts in water supply to the Roodasht irrigation network (see Fig. 11.4).

Decisions on water distribution in the Zayandeh Rud basin are made by a complex interaction of decision-making bodies on different levels based on available water stocked in the Zayandeh Rud dam as well as water rights and other criteria (for a detailed description see Chap. 3 in this volume). Interviews with farmers in Roodasht found that decision-making on water distribution is highly unclear to them, leading to great uncertainty regarding future water availability for agriculture. A notion of being deprived of water rights illegitimately consequently leads to a perception of exclusion and fraud.

Figure 11.5 shows the amount of available water in the Zayandeh Rud dam and water diverted to the Roodasht irrigation network between 2006 and 2015. Parallel to a downward trend of the water stock in the dam, surface water division to the

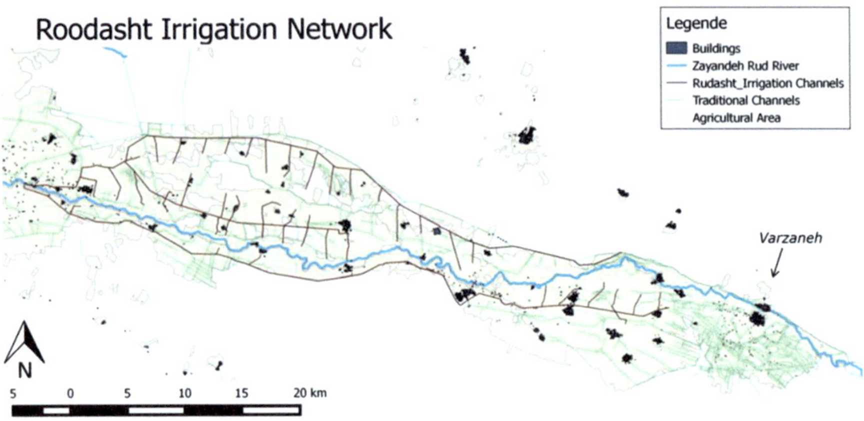

Fig. 11.4 Irrigation network Roodasht area

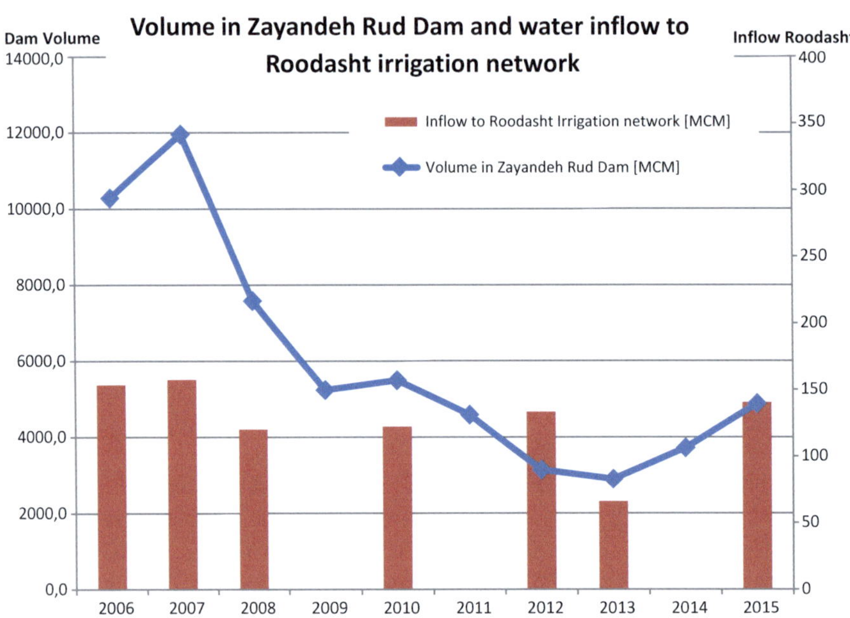

Fig. 11.5 Time series on annual water stock in the Zayandeh Rud dam and water supply to Roodasht irrigation network (data source: Isfahan Water Board Company). (A water year is presented with figures from September of the last year, until September of the given year. For example for a figure of the year 2015, data from September to December 2014 and January–September 2015 are presented)

Roodasht irrigation network is decreasing and shows strong fluctuations with several dry years with no water supply at all. In normal years water is supplied to Roodasht twice a year (in spring and autumn).

Most farmers in Roodasht use water from the irrigation network combined with other water resources which allows for farming for a period of time independently of surface water availability. Other water sources are wells, a few qanats[3] and springs. For the hydrological unit (Kouhapayeh-Segzi plain) where the Roodasht irrigation network and parts of the Abshar irrigation network are located, data from 2006/2007 suggest that almost 99% of exploited "other water resources" originate from approx. 8000 semi-deep/shallow and few deep wells.[4]

Data from the Isfahan Water Board Company also show that the yield of wells decreased by around 42% even though the number of wells increased by 11% between 2006 and 2012. The limited extraction volume may be due to declining groundwater quantity and quality in terms of salinization (with extreme EC values of up to 20 dS/m), which makes wells partly unproductive for agricultural activities. Degrading groundwater resources were also bemoaned during farmer interviews.

According to experts, increasing ground water salinity during the past decade is mainly caused by dropping groundwater tables. Limited water supply to the Roodasht irrigation network and the downstream region of the Zayandeh Rud river reduces recharge of particularly shallow aquifers and fosters overexploitation of these. Figure 11.6 shows an approx. 2 m decrease of groundwater tables of five shallow wells within 7 years between 2006 and 2013, with the most significant drops in periods of cut irrigation water in Roodasht. The figure highlights the sensitivity of shallow groundwater to surface water supply.

The complexity of problems around water availability in Roodasht is the main root cause for the thematic fields of vulnerability discussed in this chapter.

11.3.3 Agricultural Production

11.3.3.1 Soil Quality

Long-term degrading soil quality of farmland in Roodasht is a threat to high agricultural potential and fosters desertification. The main parts of agricultural soil in Roodasht have high clay content, little organic carbon and are saline/alkaline. Particularly when soils are dried out, these conditions imply a high risk of reduced productivity of soil and erosion, fostering the process of desertification.

Reduced availability of surface water forces farmers to irrigate excessively with saline groundwater to sustain their production which inevitably leads to accumulation of salts in the soil (experts report salt concentrations with EC of up to 14 dS/m). Soil degradation by salinization/alkalization has an adverse effect on agricultural yields (see Fig. 11.7) and may lead to abandoning or pausing agricultural activities

[3]Traditional underground channel with a series of vertical access shafts, used to transport water from an aquifer under a hill.

[4]Additional to these official wells, experts expect high numbers of illegal wells in Roodasht.

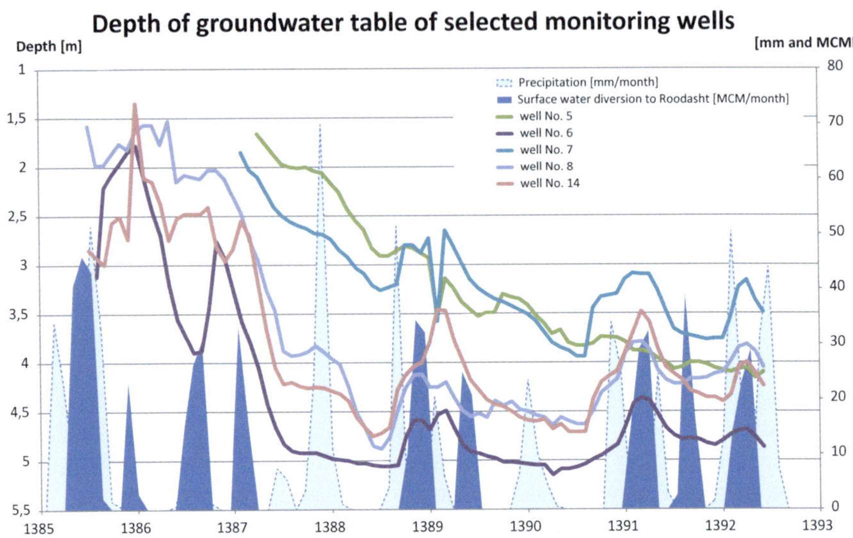

Fig. 11.6 Depth of selected monitoring wells in Roodasht with data on water supply and precipitation (data source: Isfahan Water Board Company)

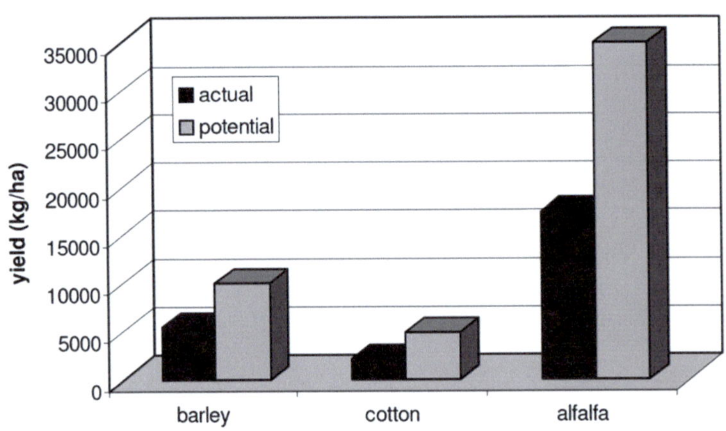

Fig. 11.7 Yield gap of selected crops due to soil and water salinization in Roodasht (source: Salemi et al. 2000)

in dry years, which increases the risk of desertification and further loss of productivity of soils.

Salinization and reduced productivity of the soil is aggravated by unfavourable agricultural practices. In general, sustainable conservation agriculture with minimal soil disturbance, permanent soil cover and crop rotations are only practiced in rare exceptions in Roodasht due to limitations in knowledge and skills as well as access to adapted farming equipment.

Regarding biomass management, the main part of the crop residuals, is removed from the field and used or sold as livestock fodder. About 1/3 of the farmers interviewed use manure as additional fertilizer but state that high costs and limited availability are practical restrictions for application, even though they are aware of the potentially positive effects on their fields.

All farmers interviewed use mineral fertilizer and pesticides/herbicides on their fields to sustain productivity. Experts note that fertilizer application is not very efficient, due to limited knowledge on soil-water-plant relationships and has the potential to further increase salinity of soils.

11.3.3.2 Cultivated Area

Next to long term effects on soil quality, cuts in surface water supply to the Roodasht irrigation network have a direct effect on the extent of cultivated areas. Figure 11.8 shows the cultivated areas in Roodasht split into areas being harvested (black column) and areas being planted but not harvested (black and white column) for the years between 2006 and 2015. The figure also shows the annual water supply (orange line) to the Roodasht irrigation network with a percentage of volume of annual water being supplied in spring time (February to July). Cultivated areas derived from remote sensing data[5] analysis are also included in the figure.

The figure shows the direct relationship of water supply and cultivated areas in Roodasht. A good proportion of cultivated areas could not be sustained and were not harvested in 2008 when surface water supply to the Roodasht irrigation network started to decrease. A reason for this may have been the decreased proportion of water being supplied during spring time (84% in 2007 and 60% in 2008) which is essential particularly for winter crops like wheat and barley (farmers have enough water to plant but not to sustain their crops over the year). This effect can also be observed in 2010 and 2012.

In the years 2009, 2011, and 2014 some cultivated areas could be sustained even without water from the irrigation network. The dataset indicates that these areas were irrigated by groundwater only and are presumably connected to (deep) wells in certain locations taping a productive aquifer resilient to limited recharge. Nevertheless a decreasing trend in water scarce years in these areas can be observed.

Interestingly in the years 2010, 2012, and 2015, the level of cultivated areas decreased by around 40% in comparison to the years 2006–2008, where similar amounts of surface water were available in Roodasht. According to experts, the reason for this phenomenon might be a strong increase in salinity and decline of (shallow) groundwater resources after dry years, which takes time to restore its capacity when surface water is available again. These restrictions for conjunctive water use leads to limitations of cultivated areas. Furthermore, agricultural areas

[5]Annual cultivated areas from the remote sensing mission (Landsat Program) have been derived by merging detected cultivated areas from the months April, May and August for each year.

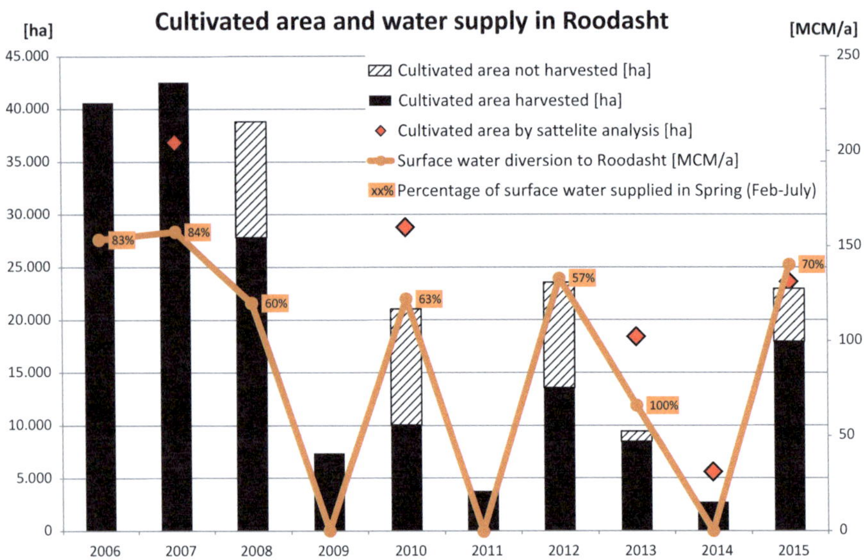

Fig. 11.8 Time series, cultivated areas merged with data on water distribution to Roodasht irrigation network and remote sensing data (data sources: AOI and Isfahan Water Board Company). (An agricultural and water year runs from approx. 20.September-20.September and is named according to the year where the calculation ends. For example: 20.9.2014–20.9.2015 is presented for the year 2015)

which are not irrigated frequently by surface water are expected to have an increase in soil salinization, making them prone to degradation and desertification particularly in desert border regions. It is assumed that these areas have been abandoned during dry years and are not easily rehabilitated to agricultural production during times when irrigation water is available. Cultivated areas in these border regions abandoned after 2007 can be observed for the downstream region of Roodasht. Figure 11.9 shows remote sensing images of cultivated areas in Roodasht for the years 2007, 2010, and 2015 are presented.

Another reason for low production in Roodasht is the common presence of small farms that are fragmented to different plots, which is typical for Iran (Ahmadpour et al. 2013). Kalantari and Abdollahzadeh (2008) and Soltani (1978) point out that the size and fragmentation of plots is an important factor for yields and production costs. Arsalanbod and Esmailpour (2000) show that for wheat production both a 1% increase in farm size or a 1% decrease in fragmentation of plots leads to a 0.4% decrease in production cost. Despite governmental subsidies, the uptake of land defragmentation measures is generally low in Roodasht since it requires cooperation between farmers, establishment of a cooperative company as well as a certain spread of know-how and investment capacity of farmers' households.

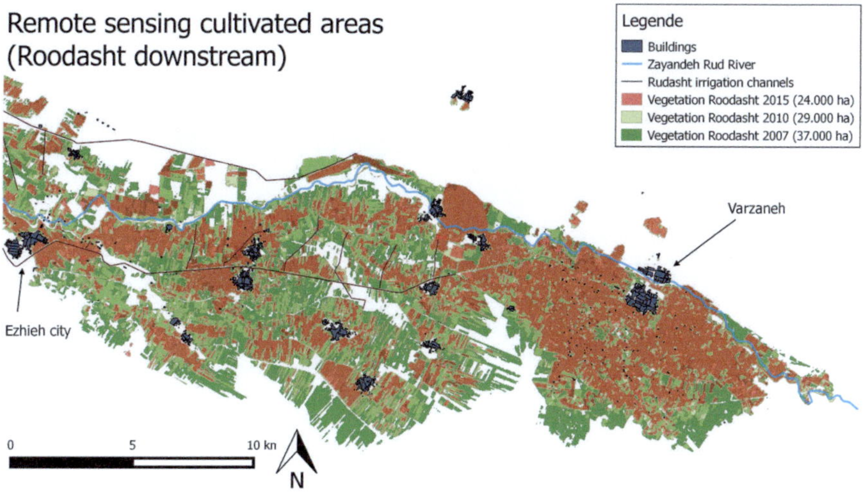

Fig. 11.9 Downstream part of Roodasht irrigation network with areas cultivated in 2007, 2010, and 2015 detected by remote sensing (data source: Technical University of Berlin, Hengsbach)

11.3.3.3 Agricultural Products

The choice of agricultural products is strongly dependent on local biophysical and infrastructure conditions, but determines income generation, options for local entrepreneurship and resilience of farmers towards water scarce conditions.

Crop choice in Roodasht is limited and dominated by (winter) wheat and barley, covering around 80% of the planted area. These grains are relatively salt tolerant, but have comparably long cultivation periods which make them dependent on irrigation water for several months, implying a high vulnerability to cuts in irrigation water particularly in spring time.

Most farmers interviewed with access to deep wells stated that they also plant other crops like alfalfa, cotton, beet, safflower, lettuce or millet and broom sorghum or were even thinking about setting up an orchard with pistachios or pomegranates.

Data on cropping patterns in Roodasht confirm the limited crop choice on fields with access to surface water and shallow wells in comparison to farms with access to deep wells or very productive aquifers (Fig. 11.10).

Figure 11.11 shows the cultivated areas for different crop types between 2006 and 2015 as well as the annual water supply to Roodasht (orange line). The proportion of wheat and barley of the total cultivated and harvested area was around 75% between 2006 and 2008. In water scarce years with low overall production this percentage decreases to 50% as more diverse production with groundwater irrigation occurs. In 2015 wheat and barley accounted for 90% of the cultivated areas. It seems these grains are the "safe option" particularly for farmers relying on surface water and shallow wells.

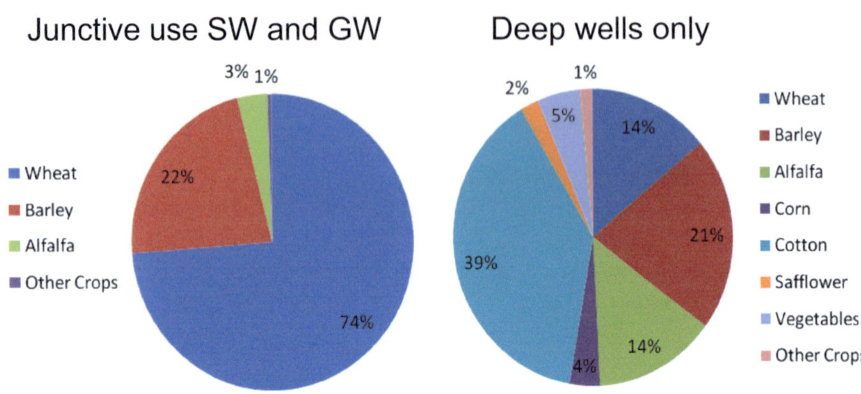

Fig. 11.10 Percentage of cultivated crop types [ha] for Bonroud district in 2015, split into areas with surface water and shallow well access only and areas with access to deep wells (source: Agricultural Organization Isfahan)

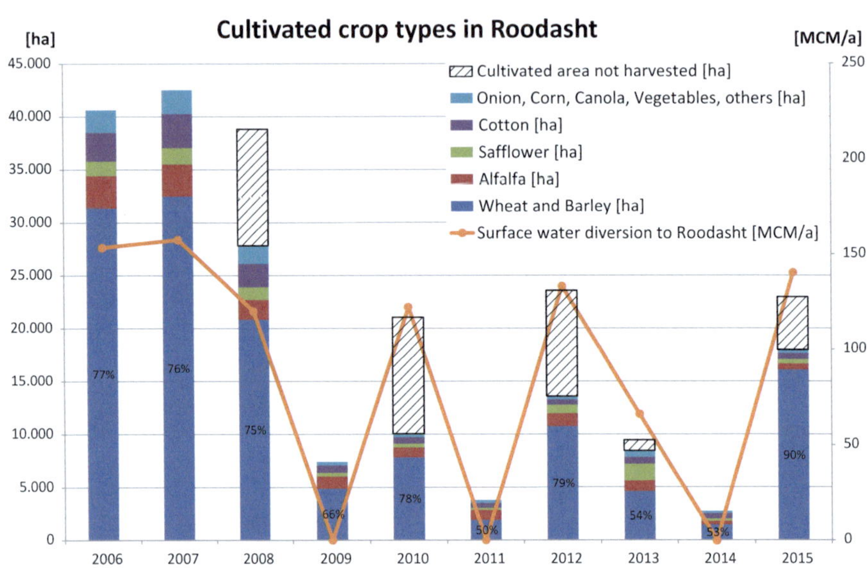

Fig. 11.11 Time series, cultivated crop types, merged with data on water distribution to Roodasht irrigation network. *Percent figures* show the proportion of wheat and barley of the correlating column (sources: AOI and Isfahan Water Board Company) (An agricultural and water year runs from approx. 20.September-20.September and is named according to the year where the calculation ends. For example: Data on water supply and cultivation between 20.9.2014–20.9.2015 are presented for the year 2015)

With regards to yields, wheat and barley are dominant in the region, followed by onion, corn and vegetables. Data show that yield per ha have not changed significantly over the years.

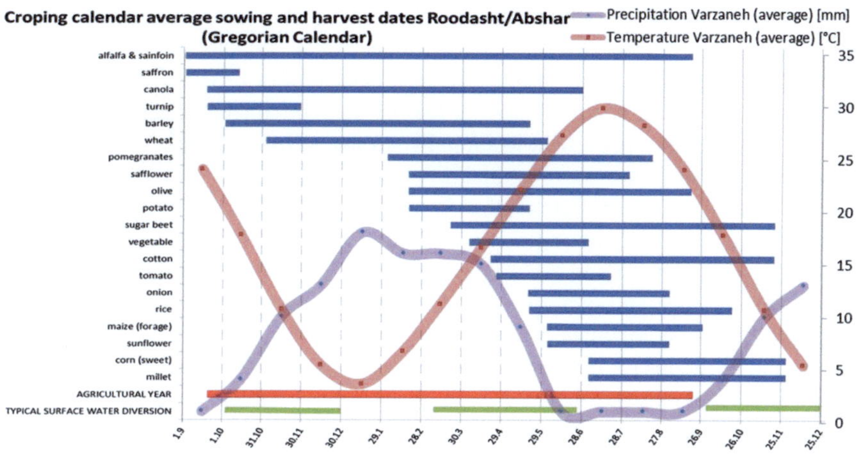

Fig. 11.12 Crop calendar for Roodasht and Abshar region, with the time period of an agricultural year and typical periods of surface water diversion to Roodasht irrigation network and average precipitation and temperatures in Varzaneh (source: Agricultural Organization Isfahan, Isfahan Water Board Company, climate-data.org)

Figure 11.12 presents an average cropping calendar for the Roodasht and Abshar regions. Time periods of an agricultural year and of typical surface water diversion to the Roodasht irrigation network are included. The figure also shows average precipitation and temperature data, related to the time scale. Comparing irrigation water availability and possible crops, it becomes obvious that timing of irrigation water does not support the cultivation of certain crops (e.g. saffron, corn, millet, etc.). It also shows that a large proportion of possible crops would be cultivated during the hottest period of the year (June to October) with almost zero precipitation, and typically no irrigation water supply to Roodasht's irrigation network. Due to an estimated irrigation water demand[6] for wheat and barley between November and May with a peak between March and May, the timing of surface water supply is optimal for these crops.

Crop choice of farmers is not very diverse in the region. During interviews, all farmers stated that the choice for wheat and barley is mainly due to low water demand and stress tolerance to drought and saline conditions, marketing at guaranteed prices with cooperative companies, production of own seeds, tradition and household consumption.

In general, there is only limited availability of improved seeds, markets for alternative products, adapted equipment but also know-how and motivation of farmers for cultivating alternative crops. Only few alternative crops like chamomile, castor seed and dill are planted on experimental farms of the Agricultural Research Centre and Agricultural Service Centres, which are used for presentation and training or by few pioneer farmers.

[6]Calculated Evapotranspiration in Varzanhe Plain with NETWAT software.

11.3.3.4 On-Farm Modernization

On-farm modernization to water-efficient agricultural production systems in Roodasht is restricted due to a range of factors. In general technical modernization is constrained due to perceived irrationality to invest in the farming sector due to high insecurity of future water availability, limited household investment capacity and little motivation to change the farming system intensively for traditional reasons and a lack in knowledge and skills, particularly among old farmers. Furthermore, experts stated that due to limited investment by farmers in past dry years, the coverage and quality of required service provision by consultancy and technical service companies has decreased. Hence, despite several governmental efforts and programs for on-farm modernization, traditional agricultural production methods with high water demand, high dependency on surface water and limited crop choice have hardly changed.

11.3.4 Farmers' Livelihoods

Knowledge of farmer households is focused on farming, deeply rooted in everyday practices, experience and skills acquired over time. Know-how is not necessarily related to formal education, as most of the interviewees have only an elementary school education. New information on farming aspects are acquired from Agricultural Service Centres and consulting companies through attending training or bilateral talks with pioneer farmers. Media, such as TV, radio or in limited cases the internet, adds to social and religious networks being used for obtaining information on farming and non-farming aspects. However, only young farmers perceive themselves as capable and willing to change and adapt their agricultural practice or start a different type of income generating activity.

But, options for income diversification through working in non-farm occupations as in local manufactories, processing facilities, industry, mines or tourism are rare in Roodasht. Farmers who pursue non-farming activities usually work as builders or drivers in the cities. Nevertheless these jobs are mainly seasonal occupations and imply low long-term income stability.

The livelihood of 75% of the farmers interviewed is entirely dependent on income generating farming.[7] For farmers with alternative sources of income, farming still accounts for more than 50% of the household income. Consequently, during water scarcity and when they could not farm, they suffered from significant cuts in their household income. There are insurances for crop loss and social security, but farmers stated that these hardly cover living costs. As one third of the interviewees own livestock, this can be an important asset during times of

[7]Farmers state to earn gross around 60–180 M IRR (1.500–4.500€) per ha a year when cultivation is possible.

low agricultural production. However, due to increased production cost and dropping prices in an unstable livestock market, many livestock keepers have sold their whole livestock below value.

The focus on traditional farming in combination with the cut in agricultural production caused significant reductions of income for farmers' livelihoods from 2009 onwards. In this situation, farmers are highly dependent on governmental financial support and loans which have piled up over successive dry years to become a huge financial burden. Continuous low agricultural incomes, financial burden of loans and debts as well as heavy inflation, make farmers prone to poverty and cause a lack of investment capacity to fund modern, adapted agricultural production systems or alternative non-farming activities. This tense situation and the risk of poverty provokes the abandoning of agriculture in order to gain income through selling land and equipment, reduce running costs for sustaining agricultural lands, or reduce the dependency on water availability. This situation makes young people in particular migrate into cities. Migration of larger numbers of farmers to cities entails the risk of segregation of this poor and poorly educated group in slums, leading to social tensions with vast impacts on cities as seen in many examples around the world.

11.3.5 Regional Entrepreneurship

Particularly in rural areas, poor access to production means like energy, water and gas infrastructure hinders local development potential for industry. This holds particularly true for processing industries (commodity chains) connected to the agricultural sector where uncertainty on future water availability makes large investments of entrepreneurs risky and irrational.

Commodity chains in the form of processing and marketing facilities for agricultural products could also be implemented collaboratively by farmers by way of cooperative companies. This type of farmers' entrepreneurship was founded in some cases particularly with livestock keepers. Establishing farmer-owned processing facilities requires know-how and a powerful collaborative initiative by farmers, which is rare in Roodasht, as in most other parts of Iran. The main reasons are mistrust amongst the members, inefficient management, a lack of experts, the lack of adequate relationships and networks between governmental organizations and rural communities, the absence of necessary equipment and facilities, and the lack of credits and vital agricultural inputs (Azkia 1992; Azkia and Ghafari 2013). Figures on the whole Zayandeh Rud basin show that with less than 6%, agricultural exploitations with cooperative companies are rather low compared to 94% individual farmers operations (Yekom 2012).

Moreover farmers' households need to dispose of a certain investment capacity to set up a business, which has diminished during dry years. To access a large share of the governmental support programs a functional cooperative structure is required. Cooperative management in terms of shared investment and joint action is a way for farmers to reduce production costs and increase agricultural productivity and income generation. In the form of NGOs and later as syndicates or cooperative companies, they could also develop alternative non-farming related activities like local (eco-) tourism. But touristic development requires an intact ecosystem, particularly in the Gavkhuni wetland as one of the regional highlights which has degraded in the last decade.

The limited regional entrepreneurship is the main reason for poor local market demand for regional (raw) products as well as little alternative employment options for farmers.

11.3.6 Environmental Conservation

Environmental degradation through desertification is an ongoing process in Roodasht. As shown in Fig. 11.13, the land cover and environment around the Roodasht irrigation network is dominated by desert and poorly vegetated range lands. Water scarcity, loss of natural ground cover, limited agricultural activities, strong winds and dust storms as well as salinization/alkalinisation of soils are drivers of an ongoing desertification process with rapid land degradation (Barzani and Salleh 2015). The Gavkhuni wetland, the mouth of the Zayandeh Rud river, is

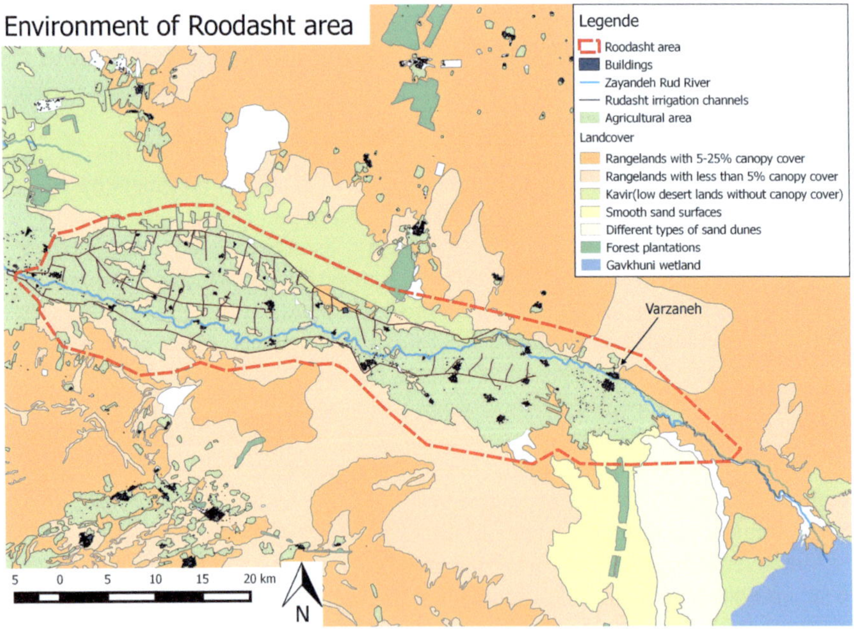

Fig. 11.13 Land cover around Roodasht irrigation network

drying out (Iranmehr et al. 2015), deteriorating the micro climate, and fostering dust storms.

The absence of conservative agricultural practices with permanent crop cover, mulching or low tillage as well as excessive application of mineral fertilizers on agricultural fields exacerbates the situation. Initiatives for environmental conservation on private land in the form of financial incentives, bids or bans, targeted information or training on conservative i.e. alternative agriculture, could not be observed. The transformation of fertile natural areas into urban or agricultural areas, free livestock grazing as well as mining activities have also degraded the quality of remnant range land patches (Bateni et al. 2012).

Little executive control of environmental offences like damaging range lands by free grazing livestock, uncontrolled mining activities or illegal water extraction from wells or from the Zayandeh Rud river, give farmers no initiative to practice environmental conservation. A reason for little execution power on the field is also that interest groups such as environmental NGOs have little, although growing, legal embedment and support.

11.4 Conclusion

With the application of an adapted conceptual framework from Füssel (2007), a range of vulnerability factors from the socio-economic and biophysical domains have been considered. Since these conditions have not yet been targeted with powerful policy measures or action plans, two main risks could be identified which are expected to grow and lead to a crisis in the near future: desertification/ land degradation and social conflict.

A high risk of desertification is founded in soil erosion and salinization, abandoned agricultural lands, little desertification control and environmental conservation, and drying out of Gavkhuni with its central function of stabilizing the local microclimate. This may have severe impacts on agricultural activities and human settlements in the whole region, risking the regional environment to degrade and making the region uninhabitable. Long term effects on the neighbouring Abshar irrigation network and even Isfahan city, with increasing sand and dust storms and changing microclimate are possible.

A high risk for social conflicts is the short and long term effect of a feeling of farmers being deprived by decision-makers and the high risk of impoverishment. The traditional form of farming is eroding and the only options seem to be to migrate to cities or remain in poverty. This setting has already caused several, also violent, protests but may escalate in the future.

Nevertheless, any kind of external stress always triggers a reaction from farmers. Adaptive capacity can be defined as the ability of farmers to adapt to external factors like water scarcity and is determined by a range of internal factors (Smit and Wandel 2006; Füssel and Klein 2006). The aspects of adaptive capacity identified, economic resources, collaborative networks, knowledge and information, infrastructure and technology, are currently low in Roodasht, but may serve as a possible entry point to develop strategies for sustainable land use and management concepts in Roodasht.

Knowledge and information dissemination are key factors for farmers for understanding the drivers of their current vulnerability and to developing adapted solutions. Informed local people can also participate in regional planning with governmental institutions.

Improved access to economic resources e.g. by targeted funding, public private partnerships or governmental incentives might empower farmers to invest in farm modernization and water-efficient agriculture or local processing facilities and other types of regional entrepreneurship. Furthermore, financial incentives to allow farmers to cover their livelihood costs with on-farm conservation measures may boost environmental conservation and reduce regional water demand.

Support and facilitation of collaborative networks e.g. by targeted training and information campaigns may enable farmers to create powerful cooperative companies. Functional cooperative companies may lead to higher uptake of land defragmentation projects, sharing farming equipment or even establishment of cooperative commodity chains, alternative income activities like eco-tourism which creates jobs and income for farmers' livelihoods, as well as a market for alternative crops.

Focused information and training campaigns could, for example, be implemented for promoting alternative farm products to more adapted species, to enable farmers to follow other jobs and develop local business, to increase farmers' awareness on environmental conservation and show opportunities for their livelihoods through these practices.

Improved access to adapted infrastructure and technology may empower farmers to adapt their farming system to modern and water efficient farming practices. Introducing environmental technologies such as decentralized energy production could support the development of local processing industry and give opportunities for local people to combine income generation with environmental conservation.

References

Adger WN (2006) Vulnerability. Glob Environ Chang 2006(16):268–228

Ahmadpour A, Feali S, Soltani S (2013) Factors affecting farmers' resistance to adoption of land consolidation case study: paddy farmers in Mazandaran province, Iran. Int J Agric Crop Sci 5(23):2871. SABI Publications

Arsalanbod M, Esmailpour A (2000) Effect of farm smallness and fragmentation of productive units on production costs: case of irrigated wheat in West Azerbaijan. Agric Econ Dev Q J 8(2): 109–116

Azkia M (1992) An evaluation of production cooperative in the Marun catchment. Zomar Consultant Engineering, Tehran, pp 81–105

Azkia M, Ghafari G (2013) Rural development, with emphasis on Iranian rural society, 4th edn. Nashreney, Tehran. (in Farsi)

Barzani MM, Salleh KO (2015) An integrated desertification hazard assessment in Iran. Int J Appl Environ Sci 10(5):1823–1834

Bateni F, Soffianian A, Fakheran S (2012) Rangeland degradation and its impacts on water quality in Zayandehroud River Basin. In: Nejadkoorki F (ed) International conference on applied life sciences. InTech, Rijeka, pp 117–121

Eslamian S, Safavi HR, Gohari A, Sajjadi M, Raghibi V, Zareian MJ (2016) Climate change impacts on some hydrological variables in Zayandeh-Rud River Basin, Iran. Isfahan University of Technology, Isfahan

Füssel H-M (2007) Vulnerability: a generally applicable conceptual framework for climate change research. Glob Environ Chang 17:155–167

Füssel H-M, Klein RJT (2006) Climate change vulnerability assessments: an evolution of conceptual thinking. Climatic Change 75(3):301–329

Gohari A, Eslamian S, Abedi-Koupaei J, Bavani AM, Wang D, Madani K (2013) Climate change impacts on crop production in Iran's Zayandeh-Rud River Basin. Sci Total Environ 442: 405–419. Elsevier

IPCC (2001) Climate change 2001: the scientific basis. In: Houghton JT, Ding Y, Griggs DJ, Noguer M, van der Linden PJ, Dai X, Maskell K, Johnson CA (eds) Contribution of working group 1 to the third assessment report of the intergovernmental panel on climate change. Cambridge University Press, Cambridge

Iranmehr M, Pourmanafi S, Soffianian A (2015) Ecological monitoring and assessment of spatial-temporal changes in land cover with an emphasis on agricultural water consumption in Zayandeh-Rud Region. Iran J Ecohydrol 2(1)

Kalantari K, Abdollahzadeh G (2008) Factors affecting agricultural land fragmentation in Iran: a case study of Ramjerd Sub District in Fars Province. Am J Agric Biol Sci 3(1):358–363

Molle F, Ghazi I, Murray-Rust H (2009) Buying respite: Esfahan and the Zayandeh-Rud River Basin, Iran, in river basin trajectories: societies, environments and development. International Water Management Institute, Colombo

Salemi HR, Mamanpoush A, Miranzadeh M, Akbari M, Torabi M, Toomanian N (2000) Water management for sustainable irrigated agriculture in the Zayandeh Rud Basin, Esfahan Province, Iran. International Water Management Institute, Colombo

Smit B, Wandel J (2006) Adaptation, adaptive capacity and vulnerability. Glob Environ Chang 16: 282–292

Soltani GR (1978) Small farm versus large farm development in Iran. Indian J Agric Econ 33(3): 94

Turner BL, Kasperson RE, Matson PA, McCarthy JJ, Corell RW (2003) A framework for vulnerability analysis in sustainability science. Proc Natl Acad Sci 14(100):8074–8079

Voss M (2008) The vulnerable can't speak. An integrative vulnerability approach to disaster and climate change research. J Civilis 1(3):39–56

Yekom consultants (2012) Agricultural studies report. Ministry of Energy, Deputy of Water and Wastewater, vol 9

Part IV
Development of Necessary Tools for IWRM and Results

Chapter 12
Introduction: The Water Management Tool WMT

Michael Kaltofen

Water management planning requires accessing the effects of changes in water demand (i.e. effects of migration or farmland and crop management, environmental needs), water yield (i.e. ground water drawdown, climate change, desertification) and water management (water allocation rules, dam operation, water transfer) on water availability and supply. Accordingly, analyses may be performed regarding one particular or any combination of these planning aspects, whereby they have to be quantified separately. Consequently, the following suitable model software has been identified:

- SWAT, Soil and Water Assessment Tool (Arnold et al. 1998),
- FEFLOW, Finite element modelling of flow, mass and heat transport in porous and fractured media (Diersch 2014),
- MIKE Basin, decision support tool for integrated water resources analysis, planning and management of river basins (DHI 2012).

These models consider the following processes:

- natural runoff as well as precipitation and potential evaporation for irrigation management (SWAT);
- climate change impact on hydrological variables (climate modelling and subsequently SWAT);
- extractions for industry (MIKE Basin);
- extractions for municipal water demand (MIKE Basin);
- extractions for water transfer from Zayandeh Rud (MIKE Basin);
- transfer to Zayandeh Rud (MIKE Basin);
- extractions for agriculture with fixed demand like pumping (MIKE Basin);

M. Kaltofen (✉)
DHI-WASY GmbH, Berlin, Germany
e-mail: kal@dhigroup.com

S. Mohajeri, L. Horlemann (eds.), *Reviving the Dying Giant*,
DOI 10.1007/978-3-319-54922-4_12

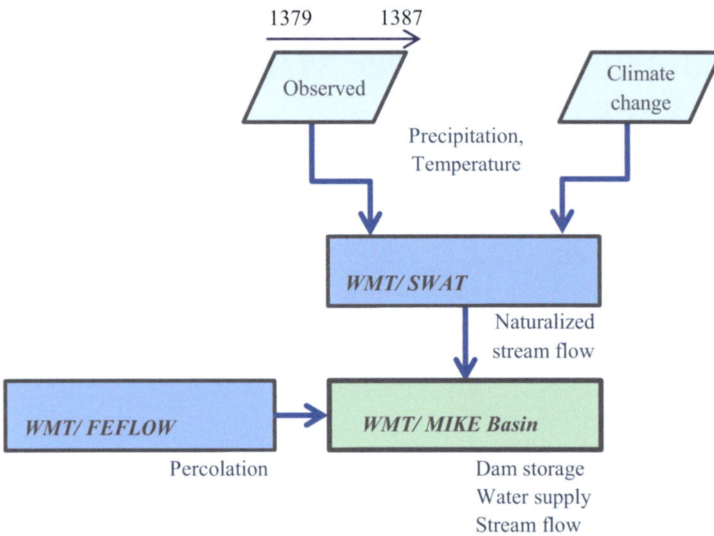

Fig. 12.1 Overview of water management tool WMT

- extractions for agriculture in irrigation areas with variable demand calculated by FAO56 approach (MIKE Basin);
- discharge by waste water treatment plants (MIKE Basin);
- seepage from Zayandeh Rud (FEFLOW, MIKE Basin);
- management of Zayandeh Rud dam (MIKE Basin); and
- water allocation among water uses (MIKE Basin).

These models and further methods and instruments for climate change analysis were coupled together in the Water Management Tool WMT (see Fig. 12.1).

The calculation of water availability in the MIKE Basin model starts from natural runoff (SWAT), before subsequently encompassing water extractions and water management such as dam regulations, water diversion and transfers as well as discharges, i.e. from waste water treatment plants are applied including percolation losses from Zayandeh Rud river (FEFLOW).

The correct (offline) interaction of the models in time and space is ensured by using the same time span (simulation period) and structure (time step) as well as the same spatial structure (i.e. hydrological catchments, river sections, irrigated areas) for transferring data.

The tasks and steps of the model development were designed according to the expectations and needs of water management planning in the Zayandeh Rud Basin. Accordingly, first the model development aimed to test the plausibility of data and rules representing the modelled processes, such as:

- precipitation and Evaporation;
- stream flow;

- cultivation and irrigation regarding the total area per network as well as the area of gardens and farmland, crop types, irrigation method and area per method and crop, water allocation rules based upon rights and shares as well as of subscriptions for each irrigation network;
- dam management;
- municipal and industrial water use and their supply level; and
- inter-basin water transfer.

For the mentioned past (time span Mehr 1379 until Shahrivar 1387), the following results were expected to be provided by the WMT:

- storage of Zayandeh Rud dam;
- amount of water supply to all users (WTP Isfahan & Yazd transfer, irrigation networks, Felman wells, etc.); and
- stream flows (i.e. Pol Zamankan, Pol Choum, etc.).

The task of the WMT development after plausibilisation was to provide the opportunity to test decisions for the time span Mehr 1379 until Shahrivar 1387 in the areas of:

- drinking water demand;
- irrigation water demand and supply from surface and groundwater;
- industry demand; and
- dam management and supply rules.

The main question that the model could answer is thus: if it had been decided in the past that . . . then the calculated impact would have happened in the past.

In the following chapters, the methods and results for each component of WMT are described in further detail. First, climate change impacts on some hydrological variables are introduced, before it is explained how meteorological data is generally transformed into naturalised stream flows. Furthermore, the ground water modelling is discussed, including the model chain pre-processing input data for MIKE Basin. Finally, insights are offered into the development and application of the MIKE Basin model.

References

Arnold JG, Srinivasan R, Muttiah RS, Williams JR (1998) Large area hydrologic modeling and assessment. Part 1: Model development. J Am Water Resour Assoc 34:73–89. doi:10.1111/j.1752-1688.1998.tb05961.x

DHI (2012) MIKE Basin – A modelling system for river basin management and planning. Model documentation – MIKE by DHI 2012

Diersch HJ (2014) FEFLOW – Finite element modeling of flow, mass and heat transport in po-rous and fractured media. Springer, Heidelberg/Berlin. doi:10.1007/978-3-642-38739-5

Chapter 13
Climate Change Impacts on Some Hydrological Variables in the Zayandeh-Rud River Basin, Iran

Saeid Eslamian, Hamid R. Safavi, Alireza Gohari, Mahdi Sajjadi, Vahid Raghibi, and Mohammad Javad Zareian

13.1 Introduction

Global warming has significant impacts on the various systems like agriculture (Bates et al. 2008), environment (Dawson and Spannagle 2009), the hydrologic cycle (Douville et al. 2002) and the urban water supply (Barnett et al. 2005; Betts 2005). Based on comprehensive reports by the Intergovernmental Panel on Climate Change (IPCC), which is done in continental scale, global warming has the effects not only on the global temperature; but also causes changes in performance of the systems which have interaction with atmosphere. Widespread impacts of climate change on the earth's climate systems have been detected by the IPCC assessments in the past periods (Fakhri et al. 2012a), including: Impact on atmosphere (e.g. change in frequency and intensity of precipitation, change in wind speed, change in atmospheric and cloud water vapour, change in the El Nino phenomenon, Monsoon, change in extreme events like flood and drought); Impact on hydrosphere (e.g. change in runoff quantity and quality, change in groundwater quantity and quality, rising global average sea level, saltwater penetration into groundwater in coastal areas); Impact on cryosphere (e.g. change in the amount of ice in the earth, change in the amount of existing glaciers, change in earth's snow cover); and impact on biosphere (e.g. change in vegetation type, change in water requirement and performance of crops, change in earth wildlife and change in the amount of erosion) (Gohari et al. 2013). Undoubtedly, human activities would increase in the future, and consequently the amount of greenhouse gases emission will increase, leading to intensified changes in global climate variables (Eslamian et al. 2011).

Atmosphere-ocean general circulation models (AOGCM) are the most reliable tools currently available for the projection of the global climate system responses

S. Eslamian • H.R. Safavi (✉) • A. Gohari • M. Sajjadi • V. Raghibi • M.J. Zareian
Department of Civil Engineering, Isfahan University of Technology, Isfahan, Iran
e-mail: hasafavi@cc.iut.ac.ir

S. Mohajeri, L. Horlemann (eds.), *Reviving the Dying Giant*,
DOI 10.1007/978-3-319-54922-4_13

under climate change (IPCC 2007). There are three type of uncertainties sources in the usage of AOGCMs at climate change impacts assessments including: the uncertainties in future greenhouse gas and aerosol emissions (Parry et al. 2004), the uncertainties in AOGCMs (Khan et al. 2006) and the uncertainties in global climate model sensitivities (Elmahdi 2008). Ignoring these uncertainties reduces the credibility of the final results and presents the unreal results for decision-makers (Prudhomme et al. 2003). Multiple climate change scenarios produced by different AOGCMs have been used to better represent and deal with uncertainties (Tao et al. 2009; Tao and Zhang 2010; Daccache et al. 2011). This study aims to evaluate the climate change impact assessment on the climate variables such as maximum temperature, minimum temperature, mean temperature and precipitation of the Zayandeh-Rud basin in Central Iran). A probabilistic ensemble model approach was used to project the climate change impacts in the study area. This ecosystem along mountain slopes is closely stacked because of sharp vertical temperature and precipitation gradients and is particularly sensitive to anthropogenic climatic change (Diaz et al. 2003; Huber et al. 2005; Bradley et al. 2006; Nogués-Bravo et al. 2007; Sorg et al. 2012). An innovate approach used to regionalize projected series in station scale to improve the reliability of the climate change impact assessment.

13.2 Zayandeh-Rud River Basin

The semi-arid Zayandeh-Rud River Basin is one of the most strategic river basins of Iran (Madani and Marino 2009). This river basin with an area of 26,917 km^2 is located in central Iran, between the 50° 24′ to 53° 24′ longitudes and 31° 11′ to 33° 42′ latitudes. The annual precipitation varies from 1500 mm in the west to 50 mm in the east of the basin. The Zayandeh-Rud River, with an average natural flow of 1400 million cubic meters (MCM) per year, including 650 MCM of natural flow and 750 MCM of inter-basin transferred flow, starts in the Zagros Mountains in the west of the basin and ends in the Gav Khuni Wetland—a preserved wetland under the Ramsar Convention—in the east of the basin. The Zayandeh Rud River is a vitally important river for agricultural development as well as domestic water supply and economic activity of the Isfahan province in west-central Iran (Safavi et al. 2015). This river passes several agricultural and urban areas, including the populous city of Esfahan—the former capital of Iran. This makes the river the most important water resource of the basin for more than 3.7 million residents and their urban, industrial, and agriculture water consumptions, as well as for the survival of the Gav Khuni Wetland and its valuable ecosystem.

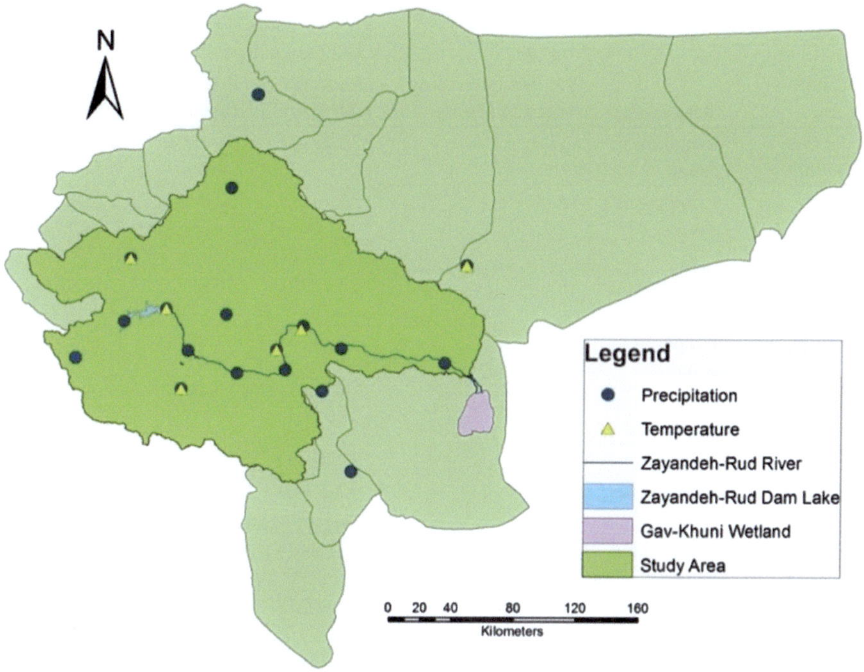

Fig. 13.1 The locations of selected gauges

13.2.1 The Station's Network in the Study Area

The time series of the climate variables are evaluated based on the time series length and the accuracy of collected climate data while the number of stations is maximized. The doubtful data are removed from the time series and then missing data are generated using WeatherMan software (Hoogenboom et al. 2005). Figure 13.1 shows the locations of the evaporation and rain gauging stations of Energy Ministry in the study area.

At the first step, 100 climatological stations of Energy Ministry are selected. As a 30 year time period (1971–2000) is selected here as a baseline period, the stations with data length less than 30 years are removed.

Therefore, 24 stations including six stations for temperature variable and 18 stations for precipitation are finally considered for this study (Fig. 13.1). Tables 13.1 and 13.2 show the characteristics of the mentioned stations.

Table 13.1 Selected rain gauges in the study area

Station name	Station type	Number of existing record (Years)	Geographic specifications Latitude (Degree)	Longitude (Degree)	Elevation (Meter)
Vezvan Meimeh	Evaporation gauge	45	33.41722	51.18917	2013
Sade Zayandeh Rud	Evaporation gauge	45	32.73417	50.7425	2173
Pol Zamankhan	Evaporation gauge	45	32.49861	50.89528	1880
Zefreh Falavarjan	Evaporation gauge	45	32.50194	51.49861	1648
Damaneh-Feridan	Evaporation gauge	45	33.01833	50.495	2388
Esfahan	Synoptic	49	32.61667	50.495	1550
Mahyar	Evaporation gauge	44	32.26889	51.80889	1686
Fin	Evaporation gauge	42	33.94444	51.37111	1050
Ghaleh Shahrokh	Evaporation gauge	41	32.66278	50.45306	2109
Pol Kaleh	Rain gauge	44	32.37306	51.23056	1771
Ziyar Bravn	Rain gauge	42	32.50611	51.94	1559
Lenj	Rain gauge	45	32.39056	51.55917	1672
Neyestanak Naein	Evaporation gauge	41	32.96806	52.80306	1910
Tiran	Evaporation gauge	45	32.70333	51.15583	1890
Varzaneh	Rain gauge	44	32.41944	52.64694	1495
Maghsoud Beyk	Evaporation gauge	43	31.81278	52.00556	1991
Chelgerd	Rain gauge	47	50.1231	32.4586	2372
Shahrekord	Synoptic	49	32.28333	50.85	2049

13.3 Method

13.3.1 Generation of Climate Change Scenarios

Global Climate Models (GCMs) are considered as the most credible tools for the projections of climate condition and extreme events in the future. In this study, the outputs of 15 GCMs under A2 and B1 emission scenarios from the Fourth Assessment Report (AR4) of Intergovernmental Panel on Climate Change (IPCC) are used (IPCC 2007). The detailed description of the selected models is provided in Table 13.3. The monthly temperature and precipitation data for the baseline period (1971–2000) and future period (2015–2044) are extracted from the Data

Table 13.2 Selected temperature stations in the study area

Station name	Station type	Number of existing record (Years)	Geographic specifications Latitude (Degree)	Longitude (Degree)	Elevation (Meter)
Neyestanak Naein	Evaporation gauge	41	32.968	52.803	1910
Sade Zayandeh Rud	Evaporation gauge	45	32.734	50.742	2173
Falavarjan	Evaporation gauge	38	32.502	51.498	1648
Shahrekord	Synoptic	49	32.283	50.85	2049
Damaneh-Faridan	Evaporation gauge	45	33.018	50.495	2388
Esfahan	Synoptic	49	32.616	50.495	1550

Table 13.3 Description of the 15 GCMs of IPCC's Fourth Assessment Report (AR4)

Model	Abbreviation	Centre
HadCm3	HADCM3	UKMO (UK)
ECHAM5-OM	MPEH5	MPI-M (Germany)
CSIRO-MK3.0	CSMK3	ABM (Australia)
GFDL-CM2.1	GFCM21	NOAA/GFDL (USA)
MRI-CGCM2.3.2	MRCGCM	MRI (Japan)
CCSM3	NCCCSM	NCAR (USA)
CNRM-CM3	CNCM3	CNRM (France)
MIROC3.2	MIMR	NIES (Japan)
IPSL-CM4	IPCM4	IPSL (France)
GISS-E-R	GIER	NASA/GISS (USA)
BCM 2.0	BCM	Beijing Climate Center (China)
CGCM3 T47	CGCM	Canadian Centre for Climate Modelling and Analysis
ECHO-G	ECHO	Meteorological Institute, University of Bonn
INMCM 3.0	INMCM	Russian Academy of Science, Institute of Numerical Mathematics
NCARPCM	NCRPCM	National Center for Atmospheric Research (NCAR), USA

Distribution Centre (DDC) of IPCC. These data are downloaded from the website (http://www.cccsn.ec.gc.ca). The differences of the temperature and relative precipitation in the 30 year monthly average future period (2015–2044) and baseline period (1971–2000) are calculated for each month.

13.3.2 The Uncertainty Evaluation of GCMs

There are high levels of uncertainty in outputs of GCMs. These uncertainties affect the confidence in the results of impact assessment studies (Fakhri et al. 2012b). There are many methods for dealing with such uncertainties, including but not limited to expression of the results as a central prediction, a central prediction with error bars, a known probability distribution function, a bounded range with no known probability distribution, and a bounded range within a larger range of unknown probabilities.

13.3.2.1 Weighting the GCMs

Each of the 15 GCMs used in this study is weighted using Eq. (13.1), based on the Mean Observed Temperature–Precipitation (MOTP) method (Massah Bavani and Morid 2005). To weight each model, this method considers the ability of that model in simulating the observed climate variables, i.e., the difference between the simulated average temperature and average precipitation in each month in the baseline period and the corresponding observed values:

$$\mathrm{Wij} = \frac{\left(\frac{1}{\Delta \mathrm{dij}}\right)}{\sum_{\mathrm{j}} \left(\frac{1}{\Delta \mathrm{dij}}\right)} \tag{13.1}$$

where, Wij is the weight of GCM j in month i; and Δdij is the difference between average temperature or precipitation simulated by GCMj in month i of base period and the corresponding observed value.

13.3.2.2 Probability Distribution of Temperature and Precipitation Changes

In this step, PDFs of climate change variables are developed for each month. These PDFs relate monthly temperature and precipitation changes to the weight of corresponding GCMs. To evaluate possible effects of climate change, discrete possibilities should be converted to continuous functions. The Beta function was then used to convert discrete probability distributions of change field scenarios to continuous probability distributions: Beta Distribution is defined by parameters of shape, upper and lower data limits. This distribution can be determined by changing shape parameter based on data's skewness. General Form of Beta distribution probability density function is:

$$f(x) = \frac{(x\text{-}a)^{p-1}\,(b\text{-}x)^{q-1}}{B(p,q)(b\text{-}a)^{p+q-1}} \qquad a << x << b; \quad p,q > 0 \tag{13.2}$$

where, p and q are the shape parameters, a and b are respectively the upper and lower data limits and B(p,q) is the Beta function. Therefore, the Beta distribution is fitted to climate change scenarios of precipitation, minimum and maximum temperature for each month.

In this part, on the basis of fitted Beta distribution functions, Cumulative Distribution Function (CDF) curve of temperature and precipitation changes of being equal to or more than the base generated for each month. Monthly ΔT and ΔP values for 50% probability level have been extracted from the CDF curves. Selected possibility level indicates an average climate condition during the future periods.

13.3.2.3 Downscaling

Low spatial resolution of models and great uncertainty in their daily outputs, especially for precipitation, result in the output that is not proper for direct use in simulating models stages and analysis of severe events. The downscaling techniques are used to bridge the gap between the GCMs' outputs and required inputs of impacts assessment models (Rajabi et al. 2010). Stochastic Weather Generator is one of statistical downscaling methods for creating daily weather scenarios in a specific station. LARS-WG is here used to generate future daily time series of maximum and minimum temperatures and precipitation under climate change scenarios.

13.3.3 Regionalization

13.3.3.1 Regionalization of Downscaled Precipitation

As the results of this study are planning to apply for the simulation of the hydrologic conditions of the Zayandeh-Rud River Basin under climate change, the climate change scenarios for 58 stations should be projected as the imputes for hydrologic model. Therefore the results of climate change in 18 stations had to be regionalized to project the climate change scenarios for all required stations (58 stations) for the hydrologic model (Fig. 13.2).

As precipitation has considerable variations in the study area, daily precipitation correlation of each of remaining stations (40 stations) with the climate change study stations (18 stations) during their available data period are calculated in order to predict the precipitation values. Four different methods including Inverse Distance Weighting, Kriging, Natural Neighbourhood and Spline are applied for regionalization. In the next step, historical daily records of four stations, named Dezak Abad

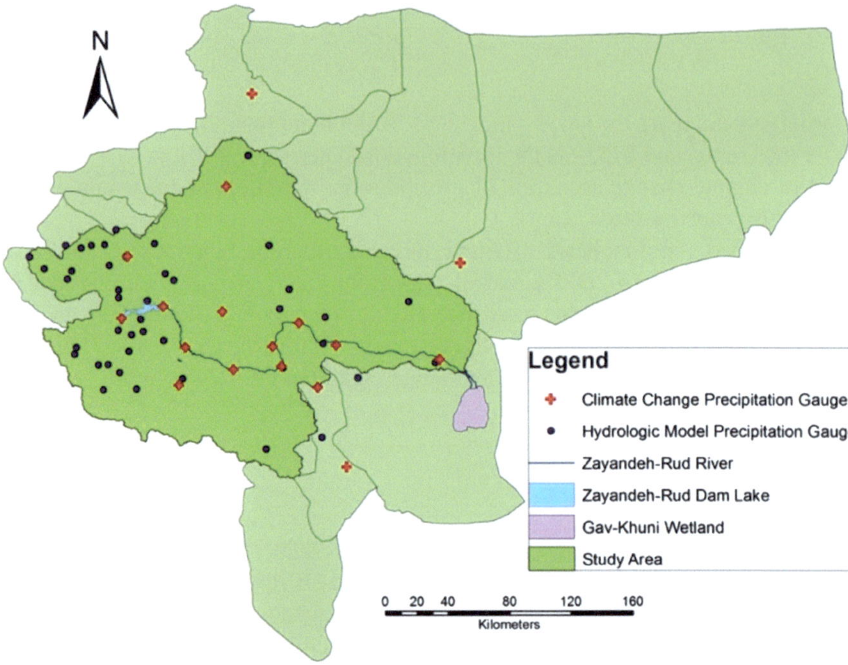

Fig. 13.2 Situation of both climate change and hydrologic model precipitation stations

and Boein located in upstream of study area and Kouhpaye and Mohammad-Abad-Jarghouye in downstream, are constructed by these four methods and compared with the observation data to compare the accuracy of the mentioned methods.

According to the better coincidence of Natural Neighbourhood predicted quantities, this method is used for the regionalization of precipitation. Comparison of the results is shown in the Table 13.4.

13.3.3.2 Regionalization of Downscaled Temperature

The six selected temperature stations is used for regionalization of temperature variables for 25 stations needed for hydrologic model. Spatial distribution of stations is shown in Fig. 13.3. Three temperature stations named Ghale Shahrokh, Kouhpaye and Morche-Khort are used to select the appropriate method for the regionalisation of temperature. Daily quantities of temperature for the three stations are interpolated by using IDW, Kriging and Natural Neighbourhood and compared with historical observations. Comparison of the results is mentioned in Table 13.5. The results showed the IDW can have acceptable capability in the regionalisation of temperature in this study.

Table 13.4 Comparison of the results of interpolation methods for aforementioned stations (in millimetres)

Station	Interpolation method	RMSE	R^2	Yearly mean precipitation (Estimated)	Yearly mean precipitation (Observed)
Boein	IDW	2.63	0.66	387	321
	Kriging	2.71	0.64	394	
	Natural Neighbourhood	2.57	0.66	371	
	SPLine	2.73	0.66	373	
Dezak Abad	IDW	5.85	0.67	1094	859
	Kriging	5.23	0.68	951	
	Natural Neighbourhood	5.64	0.67	1043	
	SPLine	5.97	0.67	1111	
Kouhpaye	IDW	1.36	0.51	143	112
	Kriging	1.36	0.51	143	
	Natural Neighbourhood	1.25	0.52	122	
	SPLine	1.42	0.5	145	
Mohamad-Abad-Jarghouye	IDW	1.23	0.59	137	96
	Kriging	1.12	0.64	131	
	Natural Neighbourhood	1.08	0.64	125	
	SPLine	1.27	0.58	138	

13.4 Results and Discussion

Figures 13.4 and 13.5 show the average of temperature changes in all weather stations for the A2 and B1 emission scenarios. These values are obtained by 30 year averaging of temperature in six weather stations for each month. Also, in this Figure, the spatial variability in temperature changes has been shown by vertical lines. The average temperature during all months is increased for all emission scenarios, though this increase is varied in different months. The maximum increase in temperature is observed in September. The minimum temperature increase occurs in January, though this is less in the initial months of the years.

Seasonal and annual changes of temperature in the Zayandeh-Rud River Basin have been summarized in Tables 13.6 and 13.7. The results show that optimistically, annual temperature in the Zayandeh-Rud River Basin will be increased by 0.6–1.1 °C in different weather stations. Summer will experience the maximum increase in temperature in all weather stations and for both emission scenarios. Winter will also have the minimum temperature increase. Overall, the A2 emission scenario shows more temperature increase in comparison with the B1 emission scenario.

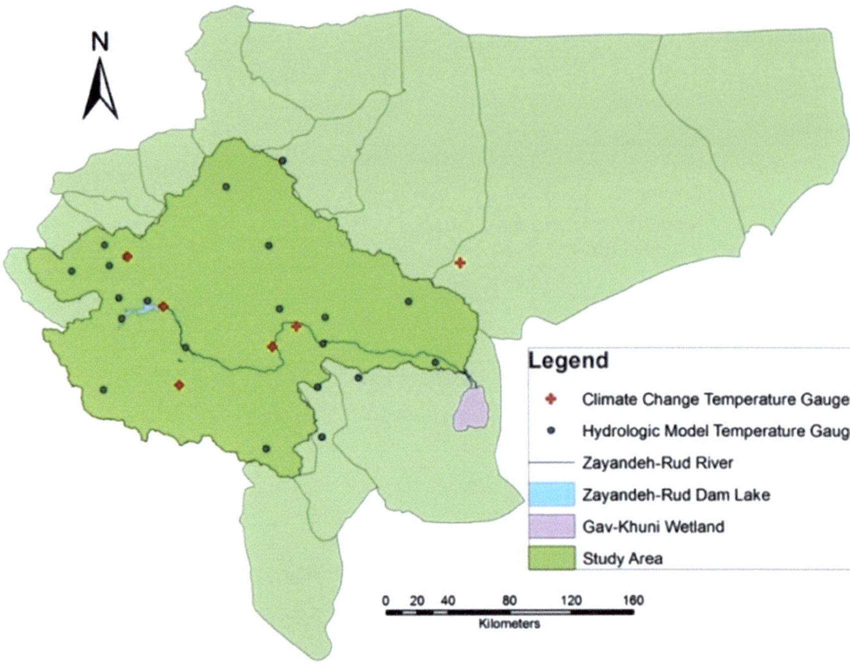

Fig. 13.3 Distribution of temperature stations for climate change and hydrologic model

Table 13.5 Comparison of the results interpolation methods for temperature in different stations

Station	Method	R2	
		Min. temp.	Max. temp.
Ghale Shahrokh	IDW	0.87	0.88
	Kriging	0.86	0.87
	Natural Neighborhood	0.87	0.88
Kouhpaye	IDW	0.9	0.93
	Kriging	0.88	0.92
	Natural Neighborhood	0.9	0.91
Morche-Khort	IDW	0.93	0.87
	Kriging	0.92	0.87
	Natural Neighborhood	0.93	0.87

Growing warming for Middle East regions is predicted by several studies using climate change indices which primarily focus on extremes (Alexander et al. 2006; Sensoy et al. 2007; Lelieveld et al. 2012). The relatively strong upward trend of temperature in the northern regions of the East Mediterranean and the Middle East indicates a continuation of the increasing intensity and duration of heat waves observed in this region since 1960 (Kuglitsch et al. 2010). Later studies also show a decrease in precipitation and increase in extreme precipitation for some

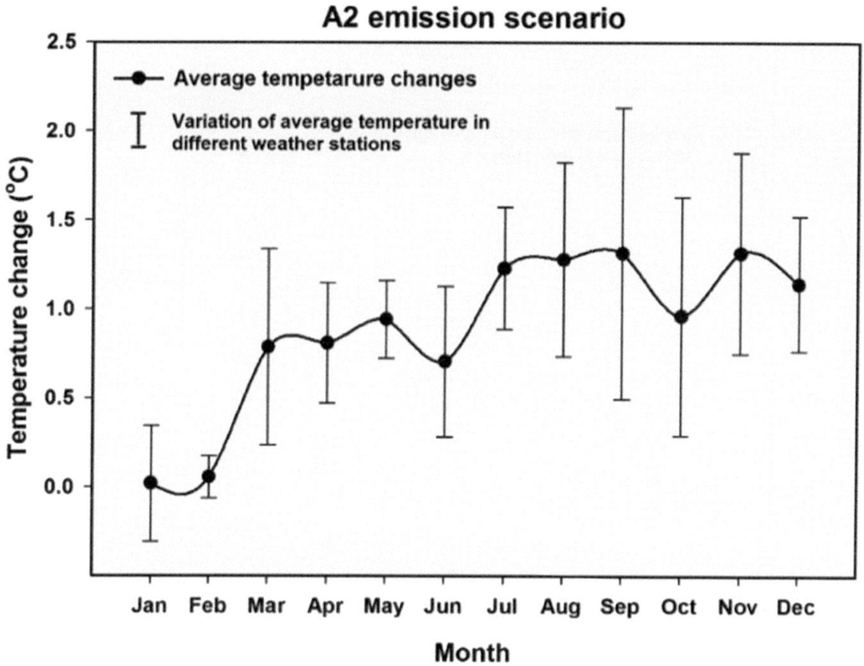

Fig. 13.4 Average monthly temperature change in different weather stations for A2 emission scenario

regions in Middle East and have a good consistency with results of current study. However, the modelled changes in precipitation exhibit a large variability in space and time and there is a less spatially coherent pattern of change and a lower level of statistical significance for precipitation compared with temperature changes.

Figures 13.6 and 13.7 show the average of precipitation changes in 18 weather stations for different emission scenarios. These values are obtained based on the 30 year averaging of precipitation predicted in each month. Also, the spatial variability of the precipitation has been shown by vertical lines. The amount of precipitation changes don't follow a uniform increasing or decreasing trend. However, in cold months (January, February and March), precipitation always decreases for A2 and B1 emission scenarios. In other hand, precipitation in future warm months is increased for all emission scenarios. However, the most precipitation in the Zayandeh-Rud River Basin occurs in the cold months.

Seasonal and annual precipitation variations in all weather stations have been shown in Table 13.8. Winter is found to have the maximum precipitation decrease in comparison with other seasons. As in 1971–2000, summer had very low precipitation, the precipitation changes rainfall in this season are negligible and had the minimum precipitation changes in A2 and B1 emission scenarios.

Comparison of annual precipitation in the near future shows that precipitation will be changed from −0.53 to +0.42% in all weather stations. Overall, the A2

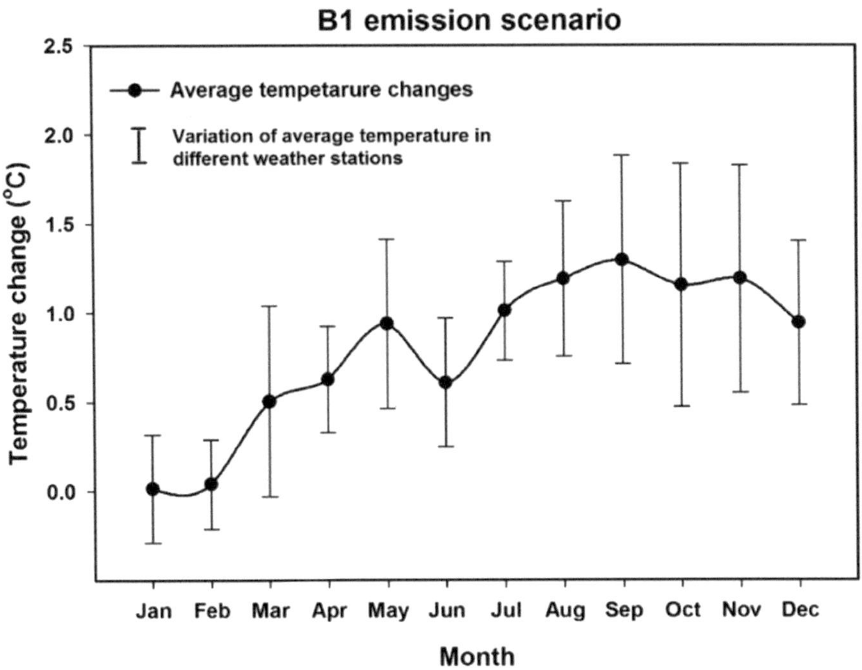

Fig. 13.5 Average monthly temperature change in different weather stations for B1 emission scenario

Table 13.6 Seasonal and annual temperature changes in different weather stations for A2 emission scenario

	Temperature changes (°C)					
Month	Neyestanak	Falavarjan	Sade Zayandeh Rud	Esfahan	Damaneh	Shahrekord
Winter	−0.2	0.4	0.5	0.4	0.4	0.3
Spring	0.3	1.0	1.0	0.8	0.7	1.0
Summer	2.0	1.1	1.4	1.4	0.6	1.0
Fall	1.9	1.1	1.3	1.2	0.6	0.9
Annual	1.0	0.9	1.1	1.0	0.6	0.8

emission scenario shows more annual precipitation decrease than the B1 emission in all weather stations.

Yearly average of baseline period precipitation, as shown in Fig. 13.8, has a high variation from the extreme value of 1450 mm/year in the west to the lowest value of 85 mm/year in the east. Meanwhile, yearly average of precipitation under A2 (B1) scenario, has predicted a decline in the region as much as 26% (27%) in the west and 7% (2%) in the east. This resulted in precipitation range fall to 1063 (1053) and 79 (83) mm/year. The gradient of precipitation changes decreased under both scenarios.

Table 13.7 Seasonal and annual temperature changes in different weather stations for B1 emission scenario

	Temperature changes (°C)					
Month	Neyestanak	Falavarjan	Sade Zayandeh-Rud	Esfahan	Damaneh	Shahrekord
Winter	−0.6	0.2	0.3	0.3	0.3	0.2
Spring	0.1	0.9	1.0	0.7	0.7	0.9
Summer	1.8	1.0	1.1	1.1	0.8	1.2
Fall	1.7	1.1	0.8	1.0	0.5	0.8
Annual	0.9	0.8	0.8	0.8	0.6	0.8

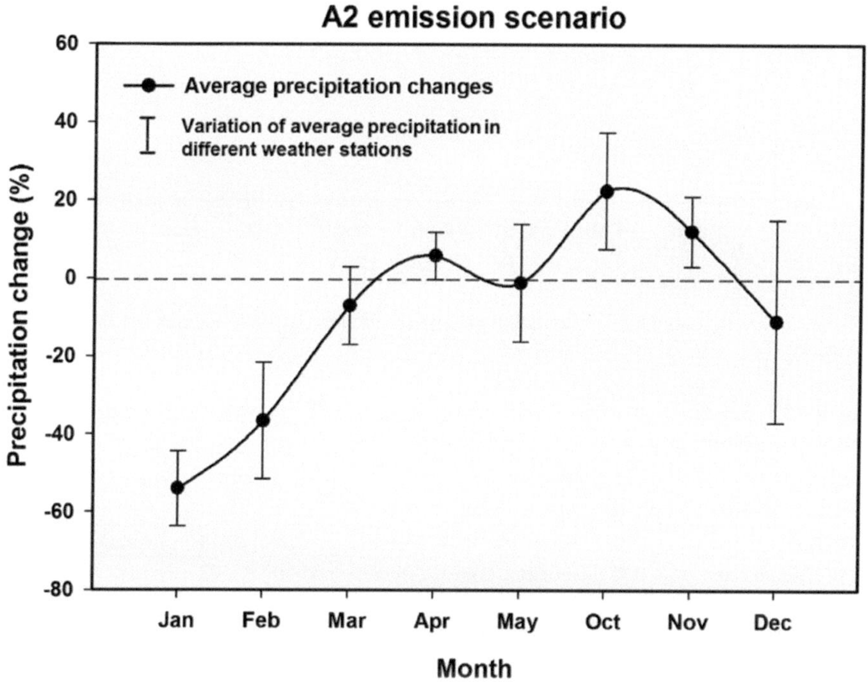

Fig. 13.6 Average monthly precipitation change in different weather stations for A2 emission scenario

Baseline mean temperature yearly average is depicted in Fig. 13.9. Low value of 9.7 °C in the west higher altitudes increases gradually to value of 16.3 °C in the east within the lowest altitudes. Under A2 (B1) scenario, it is predicted to vary temperature range of study area to 10.3–16.8 °C (10.3–16.6 °C). This means an increase of 6.2% and 3% (6.2% and 1.8%) in lower and upper bounds of temperature range, respectively. It can be seen that B1 scenario predicted intensification in gradient of variations of temperature in the study area. Central warmer region of study area has been extended to more areas under both scenarios.

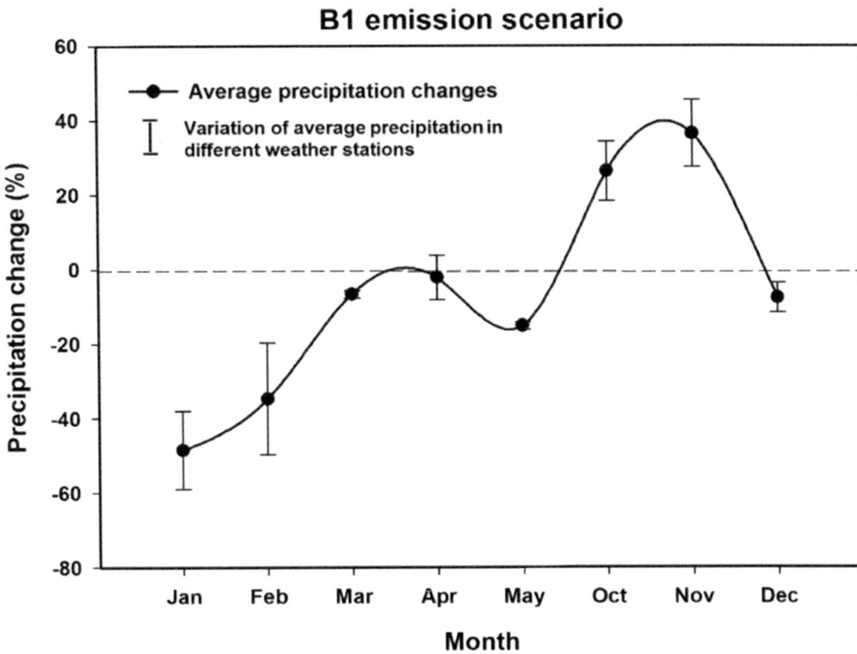

Fig. 13.7 Average monthly precipitation change in different weather stations for B1 emission scenario

Table 13.8 Seasonal and annual temperature changes in different weather stations for different emission scenarios

	A2 emission scenario				B1 emission scenario			
Weather station	Winter	Spring	Fall	Annual	Winter	Spring	Fall	Annual
Esfahan	−0.30	0.11	−0.24	−0.14	−0.26	0.14	−0.01	−0.04
Pol Zamankhan	−0.45	−0.09	0.66	0.04	−0.41	0.03	0.42	0.01
Pol Kale	−0.34	0.02	0.12	−0.07	−0.14	−0.20	0.21	−0.04
Chelgerd	−0.45	−0.03	−0.19	−0.22	−0.43	0.07	−0.23	−0.20
Damaneh	−0.31	0.16	0.13	−0.01	−0.23	0.06	0.03	−0.05
Sad Zayandeh Rud	−0.31	0.40	0.10	0.06	−0.33	−0.41	0.28	−0.15
Falavarjan	−0.50	0.06	−0.26	−0.24	−0.53	−0.30	0.28	−0.18
Ghale Shahrokh	−0.31	0.11	0.08	−0.04	−0.32	−0.05	0.25	−0.04
Lenj	−0.15	0.03	0.49	0.12	−0.18	0.01	0.30	0.04
Maghsoud Beik	−0.26	−0.18	0.18	−0.09	−0.21	−0.24	0.29	−0.05
Mahyar	−0.18	−0.19	0.09	−0.09	−0.23	−0.03	0.30	0.02
Neyestanak	−0.18	−0.19	0.09	−0.09	−0.37	0.08	−0.41	−0.23
Falavarjan	−0.51	0.06	−0.26	−0.24	−0.53	−0.29	0.28	−0.18
Shaherekord	−0.48	0.14	0.23	−0.04	−0.29	−0.05	0.27	−0.03
Fin	−0.30	−0.07	0.17	−0.07	−0.26	0.04	0.12	−0.03
Vezvan	−0.31	0.04	−0.15	−0.14	−0.22	−0.16	0.32	−0.02
Varzaneh	−0.20	0.03	0.11	−0.02	−0.16	−0.18	0.42	0.03

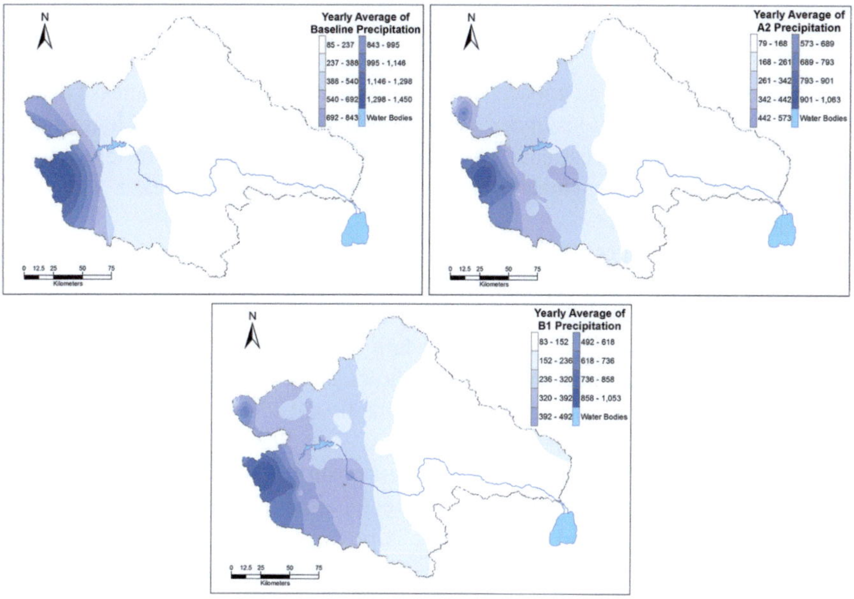

Fig. 13.8 Yearly average of baseline period, A2 and B1 precipitation

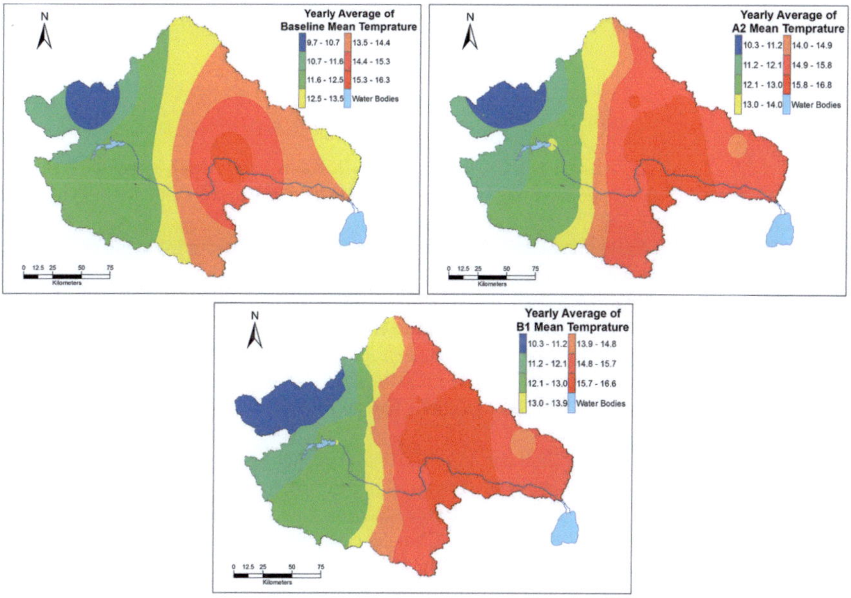

Fig. 13.9 Yearly average of baseline period, A2 and B1 mean temperature

13.5 Conclusions

Analysis of climate change impacts on the temperature indicates an increase pattern in temperature in all stations under study. General increasing trend of temperature in the upstream and downstream stations shows the lowest values of temperature increase simulated in the winter while the highest temperature increase values will be in spring and summer.

For stations in upper regions of the basin, the highest values of increase in temperature are expected in spring months. The increasing trend of winter temperature in the upper sub-basin stations can affect the amounts of snowfall and precipitation falls as rain than snow, leading to the reduction in the volume of snow pack. Generally, temperature change shows an increasing trend from the West to East and has an inverse relationship with the height.

Results of climate change impacts on the precipitation in Chelgerd station located in upstream of basin show the highest quantities of precipitation decrease at annual scale. As far as the quantities of precipitation in this region can influence considerably Zayandeh-Rud river flow, it is predicted that 25–26% reduction of annual precipitation in this area will have considerable effects on the natural river flow during the study years. The seasonal results of precipitation shows its patterns change from winter to spring and fall seasons in up- and down-stream of Zayandeh-Rud basin. In a general view, precipitation changes show a decreasing trend from the West to East and it has a direct relationship with height.

References

Alexander LV, Zhang X, Peterson TC, Caesar J, Gleason B, Klein Tank AMG, Tagipour A (2006) Global observed changes in daily climate extremes of temperature and precipitation. J Geophys Res Atmos 111(D5). doi:10.1029/2005JD006290

Barnett TP, Adam JC, Lettenmaier DP (2005) Potential impacts of a warming climate on water availability in snow-dominated regions. Nature 438(7066):303–309

Bates B, Kundzewicz ZW, Wu S, Palutikof J (2008) Climate change and water: Technical paper VI. Intergovernmental Panel on Climate Change (IPCC)

Betts RA (2005) Integrated approaches to climate–crop modelling: needs and challenges. Philos Trans R Soc Lond Ser B Biol Sci 360(1463):2049–2065

Bradley RS, Vuille M, Diaz HF, Vergara W (2006) Threats to water supplies in the tropical Andes. Science 312:1755–1756

Daccache A, Weatherhead EK, Stalham MA, Knox JW (2011) Impacts of climate change on irrigated potato production in a humid climate. Agric For Meteorol 151(12):1641–1653

Dawson B, Spannagle M (2009) The complete guide to climate change. Routledge, New York, NY

Diaz HF, Grosjean M, Graumlich L (2003) Climate variability and change in high elevation regions: past, present and future. Clim Chang 59:1–4

Douville H, Chauvin F, Planton S, Royer JF, Salas-Melia D, Tyteca S (2002) Sensitivity of the hydrological cycle to increasing amounts of greenhouse gases and aerosols. Clim Dyn 20 (1):45–68

Elmahdi A (2008) WBFS model: strategic water and food security planning on national wide level. IGU-2008 Water Sustainability Commission, Tunis

Eslamian SS, Khordadi MJ, Abedi-Koupai J (2011) Effects of variations in climatic parameters on evapotranspiration in the arid and semi-arid regions. Glob Planet Chang 78:188–194

Fakhri M, Farzaneh MR, Eslamian S, Khordadi MJ (2012a) Uncertainty assessment of downscaled rainfall: impact of climate change on the probability of flood. J Flood Eng 3(1):19–28

Fakhri M, Farzaneh MR, Eslamian SS, Khordadi MJ (2012b) Confidence interval assessment to estimate dry and wet spells under climate change in Shahrekord station, Iran, ASCE. J Hydrol Eng 18(7):911–918

Gohari A, Eslamian S, Abedi-Koupai J, Massah-Bavani A, Wang D, Madani K (2013) Climate change impacts on crop production in Iran's Zayandeh-Rud River Basin. Sci Total Environ 442:405–419

Hoogenboom G, Jones JW, Wilkens PW, Porter CH, Hunt LA, Boote KJ, Singh U, Uryasev O, Lizaso JI, Gijsman AJ, White JW, Batchelor WD, Tsuji GY (2005) Decision support system for agrotechnology transfer, Weatherman. Release 4.0.2.0. University of Hawaii, Honolulu, HI

Huber U, Reasoner M, Bugmann H (eds) (2005) Global change and mountain regions, a state of knowledge overview. Kluwer Academic Publishers, Dordrecht, 652 pp

IPCC (2007) General guidelines on the use of scenario data for climate impact and adaptation assessment. Cambridge University Press, Cambridge/New York

Khan MS, Coulibaly P, Dibike Y (2006) Uncertainty analysis of statistical downscaling methods. J Hydrol 319(1):357–382

Kuglitsch FG, Toreti A, Xoplaki E, Della Marta PM, Zerefos CS, Türkeş M, Luterbacher J (2010) Heat wave changes in the eastern Mediterranean since 1960. Geophys Res Lett 37:L04802. doi:10.1029/2009GL041841

Lelieveld J, Hadjinicolaou P, Kostopoulou E, Chenoweth J, El Maayar M, Giannakopoulos C, Xoplaki E (2012) Climate change and impacts in the Eastern Mediterranean and the Middle East. Clim Chang 114(3–4):667–687

Madani K, Marino MA (2009) System dynamics analysis for managing Iran's Zayandeh-Rud River Basin. Water Resour Manag 23:2163–2187

Massah Bavani AR, Morid S (2005) The impacts of climate change on water resources and agricultural production. J Water Resour Res 1:40–47. (In Persian)

Nogués-Bravo D, Araújo MB, Errea MP, Martinez-Rica JP (2007) Exposure of global mountain systems to climate warming during the 21st Century. Glob Environ Chang 17(3):420–428

Parry ML, Rosenzweig C, Iglesias A, Livermore M, Fischer G (2004) Effects of climate change on global food production under SRES emissions and socio-economic scenarios. Glob Environ Chang 14(1):53–67

Prudhomme C, Jakob D, Svensson C (2003) Uncertainty and climate change impact on the flood regime of small UK catchments. J Hydrol 277:1–23

Rajabi A, Sedghi H, Eslamian SS, Musavi H (2010) Comparison of Lars-WG and SDSM downscaling models in Kermanshah (Iran). Ecol Environ Conserv 16(4):1–7

Safavi HR, Golmohammadi MH, Sandoval-Solis S (2015) Expert knowledge based modeling for integrated water resources planning and management in the Zayandehrud River Basin. J Hydrol 528:773–789

Sensoy S, Peterson TC, Alexander LV, Zhang X (2007) Enhancing Middle East climate change monitoring and indexes. Bull Am Meteorol Soc 88(8):1249

Sorg A, Bolch T, Stoffel M, Solomina O, Beniston M (2012) Climate change impacts on glaciers and runoff in Tien Shan (Central Asia). Nat Clim Chang 2(10):725–731

Tao F, Zhang Z (2010) Adaptation of maize production to climate change in North China Plain: quantify the relative contributions of adaptation options. Eur J Agron 33(2):103–116

Tao F, Yokozawa M, Zhang Z (2009) Modelling the impacts of weather and climate variability on crop productivity over a large area: a new process-based model development, optimization, and uncertainties analysis. Agric For Meteorol 149(5):831–850

Chapter 14
Application of the Hydrological Model SWAT in the Zayandeh Rud Catchment

Monireh Faramarzi, Ali Asghar Besalatpour, and Michael Kaltofen

14.1 Introduction

Hydrological models are effective tools to provide framework to conceptualize and investigate the relationships between climate, human activities, and water resources (Doll et al. 2008; Bennett et al. 2013). Watershed models have been widely used in several areas including integrated watershed management, peak flow forecasting, test of the effectiveness of measures for the reduction of non-point source pollution, soil loss prediction, assessment of the effect of land use change, analysis of causes of nutrient loss, and climate change impact assessment (Tang et al. 2012). Among hydrological models, distributed models have important applications because they relate model parameters directly to physically observable land surface characteristics (Legesse et al. 2003). The Soil and Water Assessment Tool (SWAT) (Arnold et al. 1998) is a process based model, used extensively for hydrologic simulation at different spatial scales to investigate management strategies on watershed hydrology and water quality response (e.g. Schoul et al. 2008; Faramarzi et al. 2009; Faramarzi et al. 2010; Tang et al. 2012; Qiu et al. 2012; Faramarzi et al. 2013). The reliability of such applications depends on the accuracy of hydrological models in representing the physical processes, correct input data, and proper model calibration

M. Faramarzi (✉)
Department of Earth and Atmospheric Sciences, Faculty of Science, University of Alberta, Edmonton, AB T6G 2E3, Canada
e-mail: faramarz@ualberta.ca

A.A. Besalatpour
Department of Soil Science, College of Agriculture, Vali-e-Asr University of Rafsanjan, Rafsanjan, Iran

M. Kaltofen
DHI-WASY GmbH, Berlin, Germany

S. Mohajeri, L. Horlemann (eds.), *Reviving the Dying Giant*,
DOI 10.1007/978-3-319-54922-4_14

(Faramarzi et al. 2015). As such, a key challenge is initially to set up an accurate hydrological model, which correctly represents the site's actual physical processes.

With an area of about 28,000 km^2, Zayandeh Rud basin is located in central part of Iran, where hydro-climatic and geospatial conditions vary considerably from the western highlands to the eastern lowland regions. Frequent and prolonged droughts, human factors such as population growth and economic development, as well as climate change present a significant concern for reconciling the limited water resources among all conflicting sectors (i.e. energy and food production, hydro-power generation, forestry, recreation and rural development) while ensuring a sustainable economy in the basin.

We used the SWAT model in combination with Sequential Uncertainty Fitting program (SUFI2) (Abbaspour 2011) to calibrate and validate a hydrological model of the Zayandeh Rud basin using the stream flow data of 17 hydrometric stations for the 1990–2009 period. Our goal was to simulate the natural historical and future stream flow data and use it in MIKE BASIN (MB) water allocation model for water resource management of the basin. The details of the MB model have been described in an earlier chapter of this volume. Our specific objectives in this study were (1) to construct a representative hydrological model of the Zayandeh Rud basin using the best available dataset and engaging stakeholders; (2) calibrate and validate the hydrologic model of the basin using the discharge data of 17 hydrometric stations; (3) to naturalize the stream flow data as input to the MB model.

14.2 Hydrological Model

SWAT is a computationally efficient simulator of hydrology and water quality at various scales. The model is physically based rather than incorporating regression equations to describe relationships between input and output variables. It is a mechanistic time-continuous model that can handle very large watersheds in a data efficient manner and is not designed to simulate detailed single-event flood routing (Neitsch et al. 2011). Overall, the model has been developed to quantify the impact of land management practices on water, sediment and agricultural chemical yields in large complex watersheds with varying soils, land uses, and management conditions over long periods of time. The main components of SWAT are hydrology, climate, nutrient cycling, soil temperature, sediment movement, crop growth, agricultural management, and pesticide dynamics (Neitsch et al. 2011).

SWAT model spatial parameterization is performed by dividing the watershed into sub-basins based on topography. These are further subdivided into a series of hydrologic response units (HRUs), based on unique elevation, soil, land use, and slope characteristics. The responses of each HRU in terms of water and nutrient transformations and losses are determined individually, aggregated at the sub-basin level and routed to the associated reach and catchment outlet through the channel network. SWAT represents the local water balance through four storage volumes: snow, soil profile (0–2 m), shallow aquifer (2–20 m) and deep aquifer (>20 m). The soil water balance equation is the basis of hydrological modeling. The simulated processes include snow fall and snow melt, surface runoff, infiltration, evaporation, plant water uptake, lateral flow, and percolation to shallow and deep aquifers. Surface runoff is estimated by SCS curve number equation using daily precipitation data based on soil hydrologic group, land use/land cover characteristics and antecedent soil moisture. A more detailed description of the model is given by Neitsch et al. (2011). In this study, ArcSWAT 2009 was used, where ArcGIS (ver. 9.3) environment is used for project development.

14.3 Study Area

14.3.1 Climate

The Zayandeh Rud basin with an area of about 28,000 km^2 is located between 31 and 34 degrees north latitude and 49 and 53 degrees east longitude (Fig. 14.1). The Zayandeh Rud river originates from the Zagros Mountains, west of the city of Isfahan, and flows 350 kilometers eastward before ending in the Gavkhuni swamp, southeast of Isfahan city. The Zayandeh Rud river highly relies on annual snowfall in the Zagros Mountains, and therefore it is highly dependent on climate variability. The altitude varies from 1454 m to 3925 m, which has a pronounced influence on the diversity of the climate. With an average annual precipitation of 211 mm year^{-1}, the northern and high altitude areas found in the west receive about 300–1345 mm year^{-1}, while the central and eastern parts of the basin receive 75–230 mm year^{-1}. Long-term average winter temperatures of −0.4 °C in high altitudes and 11 °C in eastern low land regions, and summer temperatures of more than 7 and 24 °C in the western and eastern regions have been recorded, respectively (Fig. 14.2a–c). The temporal variations of these climate parameters are significant. Precipitation occurs mainly in winter months from December to April, and temperature reaches to 35 °C in July while it drops to −5 °C in January (Fig. 14.2d–f).

14.3.2 Hydrology

Overall, the hydrological, climatic and management conditions vary considerably in the basin. Accordingly we considered a total of seven major regions in this study to model the hydrological processes (Fig. 14.3). The main characteristics of these regions are as follows:

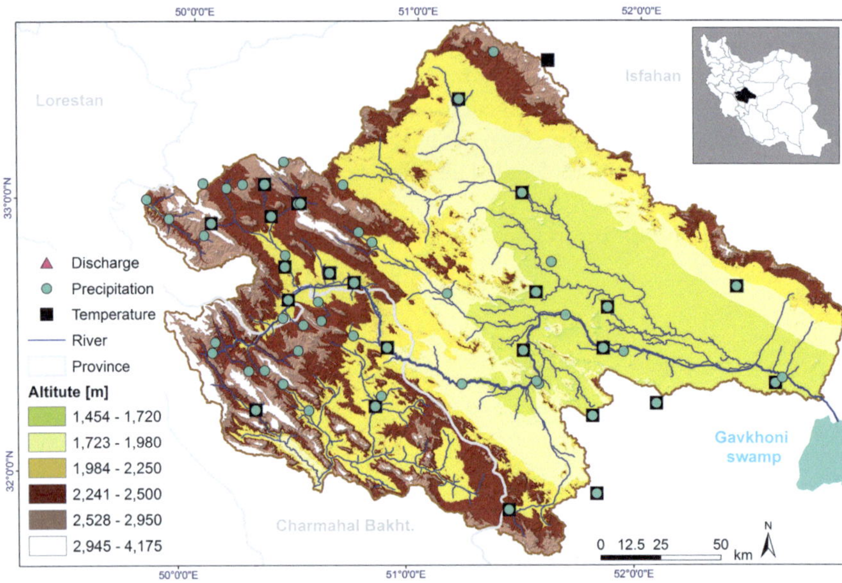

Fig. 14.1 Geographic location of the Zayandeh Rud river basin, the SRTM 90 m DEM map of the study area, and the spatial distribution of the model used precipitation, temperature, and hydrometric gauges

14.3.2.1 Upstream Zayandeh Rud Dam

A high altitude and snow fall and accumulation in fall and winter seasons and snow melt in spring time make this region different in climatic and hydrological conditions from the other regions of the study area. The precipitation is about 1400 mm in high altitude areas. With a total drainage area of about 4100 km^2, this region is the main source of water supply in the Zayandeh Rud basin. Water yield and flow regime in this region is altered mainly by water transferred from Karoon and Dez basins. The Cheshmeh Langan water transfer tunnel is being excavated for transferring water from Sardab river, Sibak river, and Cheshmelangan spring to Zayandeh Rud basin. Its yield is around 12 MCM to the basin. Koohrang Tunnels (Tunnel 1 and Tunnel 2, construction completed in 1950) redirect some of the Koohrang's water toward the Zayandeh Rud river. The annual yield of these water transfer projects are 250–300 MCM. The Sadtamzimi station is located past the Chadegan dam where the water is regulated and released for downstream users (Fig. 14.3).

14.3.2.2 Downstream Zayandeh Rud Dam

In this region the flow regime is highly regulated by the Chadegan dam, Sadtanzimi station, and later influenced by various water diversion schemes including modern and traditional irrigation networks, diversion dams alongside the river, and water

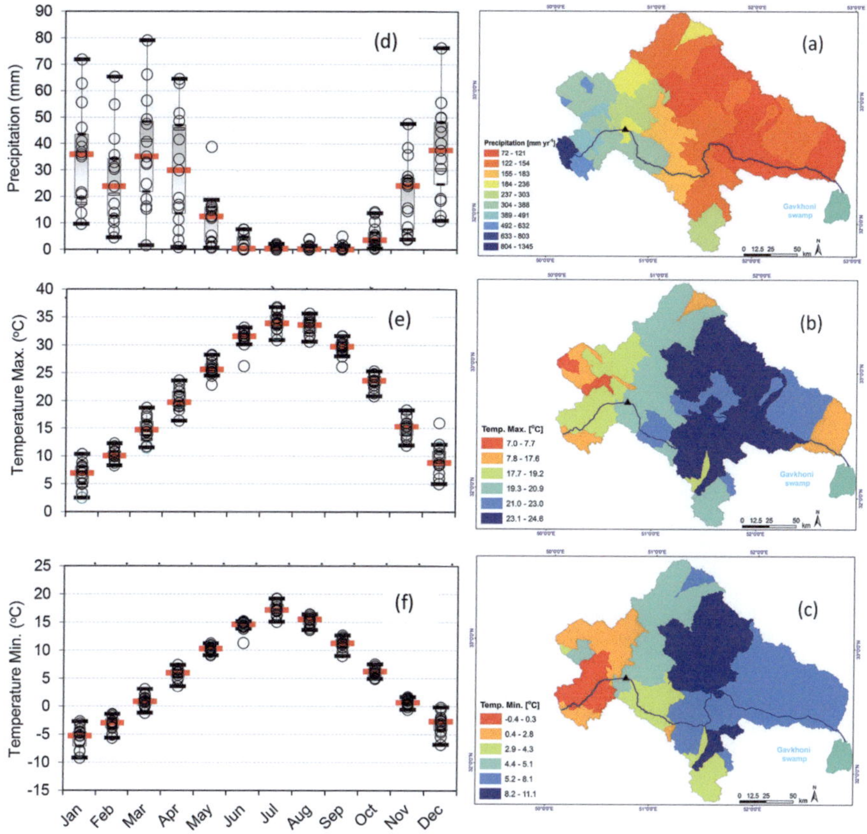

Fig. 14.2 Long-term (1990–2009) average annual (**a–c**) and monthly (**d–f**) precipitation, maximum temperature, and minimum temperature of Zayandeh Rud river basin. Circles show year to year variation of the climate variables at different months of the study period (1990–2009). Box-plots show minimum, 25th percentile, median (shown with red dashes), 75th percentile, and maximum values

extraction wells distributed on river bank (e.g. Felman wells). The natural climatic conditions vary from 300 mm year^{-1} near Chadegan dam to below 75 mm year^{-1} near Gavkhuni swamp.

14.3.2.3 Shoor River and Mobarakeh Tributary

This river originates in the southern highlands, and flows northward to end at the Zayandeh Rud right before the Dizicheh hydrometric station. The total drainage area of this river is approximately 1,653 km^2. In general the Shoor river has almost no contribution to Zayandeh Rud river. Most of the water which is generated at upstream Shoor is consumed in agricultural lands and villages in that part of the region. Therefore, the river is almost dry before it reaches to the main Zayandeh Rud river.

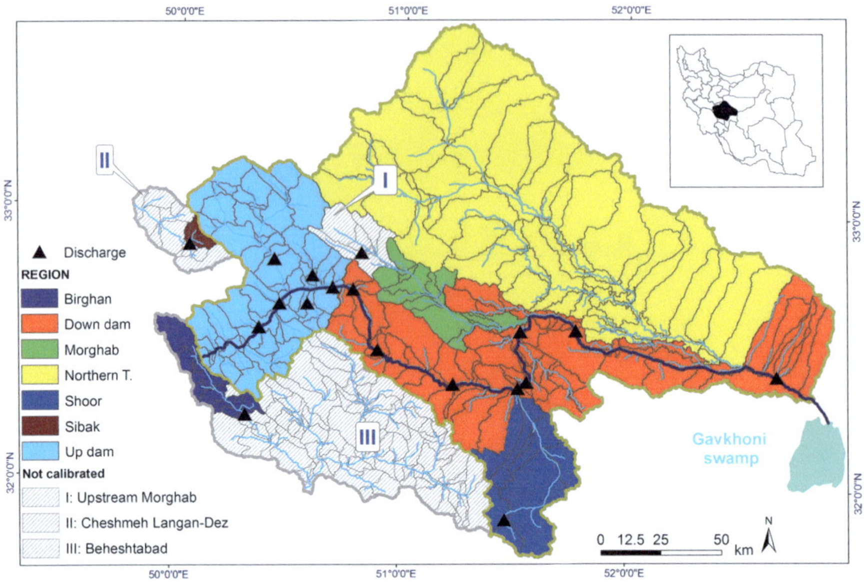

Fig. 14.3 The main hydrological regions of Zayandeh Rud basin. I: the management data are not available; II, and III: calibration of these regions is beyond the scope of our study

Similar hydrological and management conditions are to be found in Mobarakeh tributary (located west of Shoor river). The contribution of this tributary to the Zayandeh Rud is negligible.

14.3.2.4 Northern Tributaries (Dastkan River)

With a drainage area of about 12,520 km^2, the stream flow generated in this river is not reaching the Zayandeh Rud where (traditional or modern) irrigation networks are located. Similar to Shoor river basin, the discharge from tributaries in this region is fully used before it reaches to the main river or to the main irrigated networks. Water harvesting is practiced in most of the upstream tributaries during the wet seasons and water is artificially discharged into groundwater for later uses during the dry seasons. The potential water users of this region are agriculture, industries, and drinking water in rural and urban areas (Fig. 14.3).

14.3.2.5 Morghab River

Morghab river originates from streams and springs located at the northern side of Chadegan dam. Hydrological and management conditions are different in its upstream and downstream areas. Upstream, Morghab is fed mainly by the Cheshmeh-Morghab spring before the Ghalenazer hydrometric station. The

Khamiran dam is constructed in this river section after a confluence of the Cheshmeh Morghanb river and Kordolia tributary. The dam is supplied by the water yield of these two upstream tributaries and through the Karvan water transfer project. It is the source of water for downstream users e.g. Karvan irrigation networks. The total drainage area of Morghab river, from its originating tributaries to the mouth of the river where it joins the Zayandeh Rud, is about 1,463 km^2. Similar to Shoor and Dastkan rivers, stream flow generated in this river basin does not reach to Zayandeh Rud as it is used partly outside of irrigation networks.

14.3.2.6 Behestabad-Birghan River Basin

We included in this study the Beheshtabad river basin where the two main water transfer projects (Koohrang Tunnels 1 and 2) have existed for years. This river basin is not part of the Zayandeh Rud basin but is the potential source for future water development plans and belongs to the upstream catchments of the Karoon-Dez river system in western Iran. In this study, however, we opted to calibrate upstream Dezakabad hydrometric station with a drainage area of about 614 km^2. The Birghan river (upstream tributaries of Beheshtabad) flows in this basin and Koohrang dams are located on this river.

14.3.2.7 Cheshmeh Langan Spring-Sibak Basin

The Cheshmeh Langan spring is located in the upstream part of Dez river basin. A tunnel was constructed to transfer its water into Cheshmeh Langan dam located at the confluence of Sardab and Sibak rivers upstream of the Zayandeh Rud basin. The only hydrometric station measuring stream flow in this region is Charkhfalak station downstream of Sibak river. While we incorporated the upstream Dez basin in this study, we calibrated only the Sibak river, with an area of about 95 km.

14.3.3 Management

The water management and regulation of the Zayandeh Rud basin has changed over time. Before the 1960s, the distribution of water in the Zayandeh Rud watershed followed the 'Tomar', a document based on which the Zayandeh Rud stream flow was divided into 33 parts which were then specifically allotted to the eight major districts within the region. At the district level the water flow was divided either on a time basis, or by the use of variable weirs, so that the proportion could be maintained regardless of the height of the flow. After 1960, population growth, economic development within the basin, and rising standards of living particularly within the city, caused an increasing pressure on water resources so that the division of water according to Tomar was no longer feasible. Development of modern

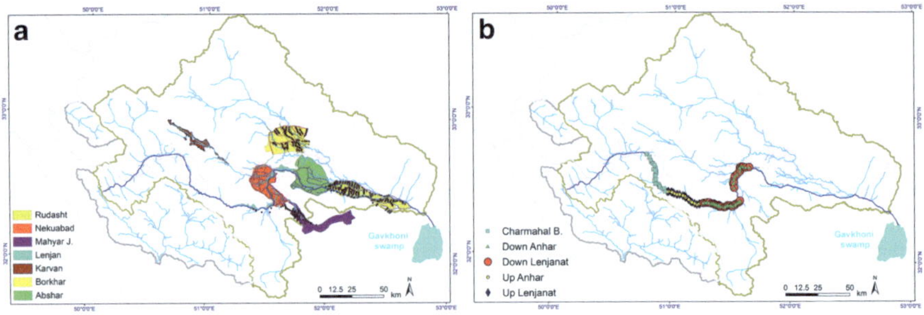

Fig. 14.4 Modern (**a**) and traditional (**b**) networks of water supply in the Zayandeh Rud river basin

irrigation networks (Fig. 14.4a) instead of traditional diversion through pumping from the shallow and deep wells (Fig. 14.4b), the creation of large steel works, Zobahan-e-Isfahan, Foolad Mobarekeh steel companies, Isfahan's petrochemical, refinery and power plants, and other new industries demanded more water than the past.

Agriculture is reported to be the largest water user accounting for about 80% of total water consumption in the basin. Three different sources of water supply the agricultural sector. These are surface water (mainly from Zayandeh Rud river), groundwater (mainly in the plains located far from the main river to access surface water), and waste water (WW). Among them the portion of surface water is the largest compared to WW which accounts for a sparse supply.

In response to increasing water demand, three main inter-basin water transfer projects were developed to convey water from the head water tributaries of Karoon-Dez river basin in south west Iran into the western upstream tributaries of Zayandeh Rud basin. These are Koohrang Tunnel 1 (Chelgerd WT project), Koohrang Tunnel 2 (Darreh Dor WT project), and Cheshmeh Langan or Vahdatabad WT project. Large hydraulic structures such as tunnels, channels, pipes, dams, and pumping stations were constructed to implement water transfer projects.

In addition to the inter basin water transfer projects, a multipurpose storage reservoir with an average annual outflow of 47.5 m^3 s^{-1} was constructed in 1972 to regulate and to allocate water during drought and dry spells. With a storage capacity of 1500 MCM, the reservoir captures most of the spring floodwater and releases it gradually throughout the summer. This has enabled the expansion of summer cropping lands (rice and maize), hydropower generation of 55.2 MW, utilization to meet the industrial and urban water demands, as well as flood control in spring time.

In this study we have considered the operation of major water management projects in the hydrological model of the basin as these anthropogenic changes can alter the downstream hydrological system. It is worth mentioning that our goal is to simulate both the Zayandeh Rud river basin and the upstream catchments of the Karoon-Dez watersheds which are considered as the major source of water supply for the Zayandeh Rud basin, through current and future water development and

Table 14.1 Input data used to build hydrological model of the study area

Data type	Data name	Time span	Resolution/ time step	Nr. of stations	Source
Climate	Precipitation (Rain gauge)	1990–2009	Daily	32	Isfahan Regional Water Board
	Precipitation (Evaporation gauge)	1990–2009	Daily	13	Isfahan Regional Water Board
	Precipitation (Climatology)	1990–2009	Daily	8	Iranian Meteorological organization
	Precipitation (Synoptic)	1990–2009	Daily	5	Iranian Meteorological organization
	Maximum and minimum temperature (Evaporation gauge)	1990–2009	Daily	11	Isfahan Regional Water Board
	Maximum and minimum temperature (Climatology)	1990–2009	Daily	9	Iranian Meteorological organization
	Maximum and minimum temperature (Synoptic)	1990–2009	Daily	5	Iranian Meteorological organization
Digital maps	DEM SRTM	2008	90 m × 90 m	–	Jarvis et al. (2008)
	Land use/Land cover	2005	1:250,000	–	Iranian Forest, Rangeland and Watershed Management Organization (IFRWMO)
	Soil map (map of land unites and map of land components)	2009	152 soil types	–	Jahad–e–Agriculture
	Soil properties	2000	maximum of 6 soil layers	353 soil profile	Isfahan Agricultural Research Institute
Management	Reservoir/dam	1990–2009	Daily	1	Iranian Ministry of energy; Isfahan Regional Water Board
	Inter-basin water transfer (Inflow to the basin: Koohrang1-2, and Cheshmeh Langan)	1990–2009	Daily	3	Iranian Ministry of energy; Isfahan Regional water Board

(continued)

Table 14.1 (continued)

Data type	Data name	Time span	Resolution/ time step	Nr. of stations	Source
	Consumptive water use: Modern Irrigation Networks	1990–2009	Monthly/ seasonal	7	Isfahan Regional Water Board
	Consumptive water use: Traditional Irrigation Networks	1990–2009	Monthly/ seasonal	Water wells of five regions along the river	Isfahan Regional Water Board
	Consumptive water use: Industry	1990–2009	Monthly/ seasonal	major industries	Isfahan Regional Water Board
	Consumptive water use: Municipal	1990–2009	Monthly/ seasonal	Isfahan city and others	Isfahan Regional Water Board
	Water transfer to other basins (Yazd and Kashan WT projects)	1990–2009	Monthly/ seasonal	2	Isfahan Regional Water Board
Hydrometric	Stream flow	1990–2009	Daily/ monthly	17	Iranian Ministry of energy; Isfahan Regional Water Board

transfer projects. Therefore, our modelled area is larger than the Zayandeh Rud basin (Figs. 14.1 and 14.3).

14.4 Input Data and Model Setup

Table 14.1 summarizes input data used to develop the SWAT hydrological model of the study area. A digital elevation model (DEM) was used for stream network and sub-basin delineation.

Figure 14.1 shows different elevation bands classified using the ESRI 90m DEM for the study area. The land use/land cover map was obtained from the Iranian Forest Rangeland and Watershed Management Organization (IFRWMO). The map has a spatial resolution of 1:250,000 and it is created using Landsat images of the year 2005 (Fig. 14.5a). With this resolution, 27 land use classes were identified in the study area. We further used SWAT database to initially characterize each land use class for the study area. A total of 41 land use parameters were assigned for each land use class illustrated in Fig. 14.5a. A soil map of basin coverage was not available. We used different components available from different organizations to build a soil map of the study area. The soil properties were acquired from Isfahan

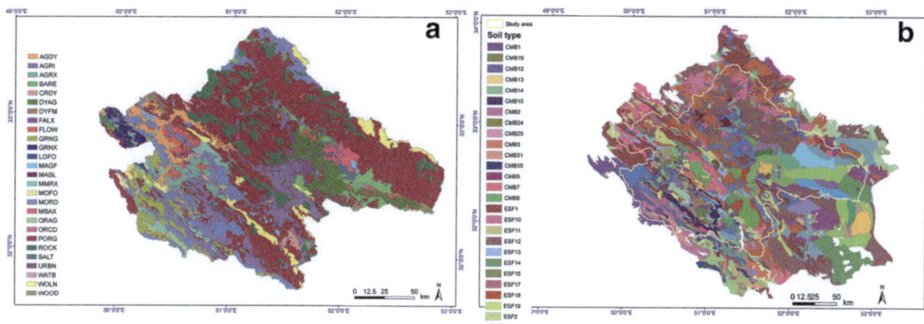

Fig. 14.5 Land use (**a**) and soil (**b**) maps of the study area. Example land use classes and soil types are shown in the legends

Agricultural Research Institute (IARI). Soil properties of approximately 353 soil profiles were studied by IARI, and data including sand, silt, and clay contents, rock fragment content, organic carbon content, soil electrical conductivity, water content, porosity, bulk density, saturated hydraulic conductivity, and soil hydrologic groups were available for each soil profile (Fig. 14.5b).

Climate data, including daily total precipitation (mm), maximum and minimum temperature (°C), and snow fall (mm.H_2O) were obtained from various sources in the study area (see Table 14.1). Other than snow fall data, the rest is required as input data for the SWAT model. Snow fall data was used for model verification in terms of partitioning the total precipitation to rain and snow. A preliminary analysis of climate data resulted in the selection of older stations which provided longer time series and contained less missing values. The availability of climate data was the main criterion to decide the study period. Figure 14.1 shows the distribution of the climate stations used in this study. Likewise, only 17 hydrometric stations were selected for the calibration and validation of the SWAT simulation results (Figs. 14.1 and 14.3).

We used the digital maps (i.e., DEM, land use, and soil) to delineate and characterize spatial units and river network for the study area. Providing a 20 km^2 as the minimum threshold area and eliminating unnecessary outlets created by the model, we delineated a total of 370 sub-basins for the study area. We considered dominant land use, dominant soil, and dominant slope for each sub-basin to balance data resolution and model complexity.

As the management control structures and water diversion plans can disrupt natural processes, we incorporated the operation of various management options in our hydrological model. These included diversion of surface water for consumptive use by various sectors, upstream flow regulation by reservoir/dam, and inter-basin water transfer projects. The consumptive water use included monthly water diversion through traditional and modern irrigation networks, and water needs of industries, and domestic sectors as well as those transferred to Yazd and Kashan provinces. A detailed analysis of water use data indicating monthly and seasonal fluctuations was conducted by the project team and was fed into the SWAT hydrological model.

A preliminary analysis of the available data for the inter-basin water transfer plans revealed that in Koohrang Tunnel 1, Koohrang Tunnel 2, and Cheshmeh Langan projects the daily data was not consistent and subject to some missing data for the earlier periods. We estimated the missing data at stakeholder meetings and using other available information. Further we calculated the monthly data and fed them into the SWAT model for the period 1990–2009 (Calibration-validation period).

We used the daily outflow of the Sadtanzimi and allowed the model to simulate the Zayandeh Rud dam's operation through modelling upstream water inflow to the reservoir and the water inflow-outflow processes in the reservoir behind the dam. It is important to mention that although the measured outflow data of dam was provided as input to the model, a proper simulation of the dam outflow in the model relies on the accuracy of the stream flow simulation in the upstream catchments. An improper stream flow simulation at the upstream tributaries of the dam can result in incorrect inflow to the reservoir which then results in emptying or overflowing of the dams (Faramarzi et al. 2015).

In this study, surface runoff was simulated using the SCS curve number method. Potential evapotranspiration (PET) was simulated using the Hargreaves method (Hargreaves and Samani 1985). Actual evapotranspiration (AET) was predicted based on the methodology developed by Ritchie (1972). The daily value of the leaf area index (LAI) was used to partition the PET into potential soil evaporation and potential plant transpiration. LAI and root development were simulated using the "crop growth" component of SWAT. This component represents the interrelation between vegetation and hydrologic balance. The calibration and validation time period was from 1990 to 1998 and 1998 to 2009, respectively, where the first three years were used as the model warm-up period.

14.5 The Calibration Program SUFI-2 and Calibration Setup

We used the SUFI-2 program to calibrate and validate the model using the observed monthly river discharge data of 17 stations for the years 1990–1998 and 1998–2009, respectively. Using this program the prediction uncertainties associated with input data (e.g., rainfall), conceptual model (e.g., process simplification), and model parameters (non-uniqueness) are aggregated and mapped into the parameter ranges. The parameter uncertainty leads to uncertainty in the output which is quantified by the 95% prediction uncertainty (95PPU) calculated at the 2.5% (L95PPU) and the 97.5% (U95PPU) levels of the cumulative distribution obtained through Latin hypercube sampling. Starting with large but physically meaningful parameter ranges that bracket 'most' of the measured data within the

95PPU, the SUFI2 decreases the parameter uncertainties through a semi-automated calibration procedure iteratively. In this procedure the best parameter set obtained through previous iteration is considered as the base to narrow the uncertainty band in the next iteration. In this iterative simulation technique where predicted output is presented by a prediction uncertainty band instead of a signal, two different indices are used to control the prediction performance (Abbaspour 2011).

Sensitivity, calibration, validation, and uncertainty analysis were performed for the hydrology using monthly river discharge data. As the SWAT model involves a large number of parameters, sensitivity analysis was essential to identify the key parameters across different regions of the study area. For the sensitivity analysis, 22 parameters integrally related to stream flow were initially selected (Table 14.2). In a second step, these parameters were further differentiated by main hydrological regions (Fig. 14.3) in order to account for regional and spatial variation in climate and management conditions. This resulted in 102 spatially scaled parameters. For this study, to better account for the regional diversity, each hydrological region was parameterized and calibrated separately.

14.6 Precipitation, Temperature, and Snow Simulation

The simulation results showed an average annual precipitation of 257 mm year^{-1} for the study area. We found that the high altitude areas found in the west received about 300–1345 mm year^{-1}, while the central and eastern parts of the basin received 75–230 mm year^{-1}. Long-term average winter temperatures of −0.4 °C in high altitudes and 11 °C in eastern low land regions, and summer temperatures of more than 7 and 24 °C in the western and eastern regions were found, respectively. The temporal variations of these climate parameters were significant. The precipitation occurred mainly in winter months from December to April, and temperature reached to over 35 °C in July while it dropped to below −5 °C in January. We also verified the simulated precipitation data with those of reported by Water and Sustainable Development Reports (WSDR). For the purpose of comparison, the weighted averages of precipitation at sub-basin level were computed and aggregated to account for the average annual (1990–2009) precipitation for every WSDR sub-catchments (Fig. 14.6). The results showed that in most of the sub-catchments simulated precipitation agrees well with those of reported by WSDR.

To check the accuracy of the model in partitioning total precipitation as snow and rain, the simulated snow fall values were compared with that of observed in highland stations (Fig. 14.7). This is important because as compare to the rain, snow fall has significant but not direct contribution to the stream flow. As shown our simulation results agreed well with the measured data.

Table 14.2 Initially selected input parameters in the calibration process

Parameter name[a]	Definition	t-value[b]	p-value[c]
v__SURLAG.bsn	Surface runoff lag time (days)	2.988	0.0003
v__SMTMP.bsn	Snow melt base temperature (°C)	7.002	1.25×10^{-9}
v__SFTMP.bsn	Snowfall temperature (°C)	5.1235	7.22×10^{-8}
v__SMFMN.bsn	Minimum melt rate for snow during the year (mm/°C-day)	3.21	0.0041
v__TIMP.bsn	Snow pack temperature lag factor	3.121	0.019
v__CH_K2.rte	Effective hydraulic conductivity in main channel alluvium (mm/hr)	2.812	0.021
r__CN2.mgt	SCS runoff curve number for moisture condition II	19.232	1×10^{-15}
v__ALPHA_BF.gw	Base flow alpha factor (days)	3.051	0.0513
v__REVAPMN.gw	Threshold depth of water in the shallow aquifer for 'revap' to occur (mm)	2.187	0.0125
v__GW_DELAY.gw	Groundwater delay time (days)	3.988	0.00021
v__GW_REVAP.gw	Groundwater revap coefficient	3.102	0.00221
v__GWQMN.gw	Threshold depth of water in the shallow aquifer required for return flow to occur (mm)	2.015	0.00621
v__RCHRG_DP.gw	Deep aquifer percolation fraction	6.184	2.71×10^{-6}
v__ESCO.hru	Soil evaporation compensation factor	6.238	1.02×10^{-9}
r__SOL_AWC.sol	Soil available water storage capacity (mm H_2O/mm soil)	9.041	3×10^{-15}
r__SOL_K.sol	Soil conductivity (mm/hr)	3.018	0.0129
r__SOL_BD.sol	Soil bulk density (g/cm^3)	8.005	2.01×10^{-13}
v__SMFMX.bsn	Maximum melt rate for snow during the year (mm/°C-day)	0.059	0.821
v__EPCO.hru	Plant uptake compensation factor	2.001	0.362
r__OV_N.hrul	Manning's n value for overland flow	0.002	0.722
r__SOL_ALB.sol	Moist soil albedo	0.302	0.933
v__CH_N2.rte	Manning's n value for main channel	0.901	0.299
v__CH_N1.sub	Manning's n value for tributary channel	1.120	0.29
v__CH_K2.rte	Effective hydraulic conductivity in the main channel (mm/hr)	1.528	0.41

[a]v__: The parameter value is replaced by given value or absolute change; r__: parameter value is multiplied by (1+ a given value) or relative change (See Faramarzi et al. 2009 for more detail)
[b]t-value indicates parameter sensitivity. The large the t-value, the more sensitive the parameter
[c]p-value indicates the significance of the t-value. The smaller the p-values, the less chance of a parameter being accidentally assigned as sensitive

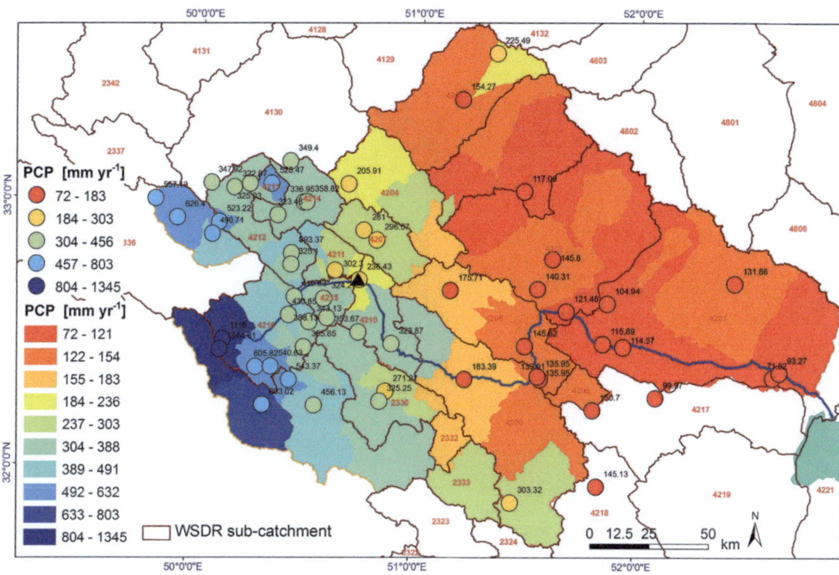

Fig. 14.6 Long-term (1990–2009) average annual precipitation. The background colors represent SWAT simulation results in each sub-basin; and the observed values are shown in circles for different rain gauges

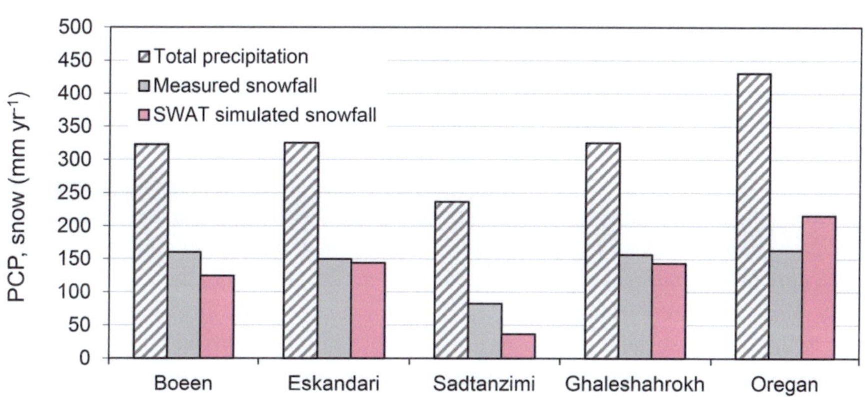

Fig. 14.7 Comparison of measured and simulated snow fall (mm) at selected high-altitude gauges

14.7 Stream Flow Simulation and Calibration-Validation Results

To calibrate and validate the hydrological model, we started with one first run to get an indication of the model performance and observed discharge stations to be used for the calibration. The results showed that many stations are highly influenced by:

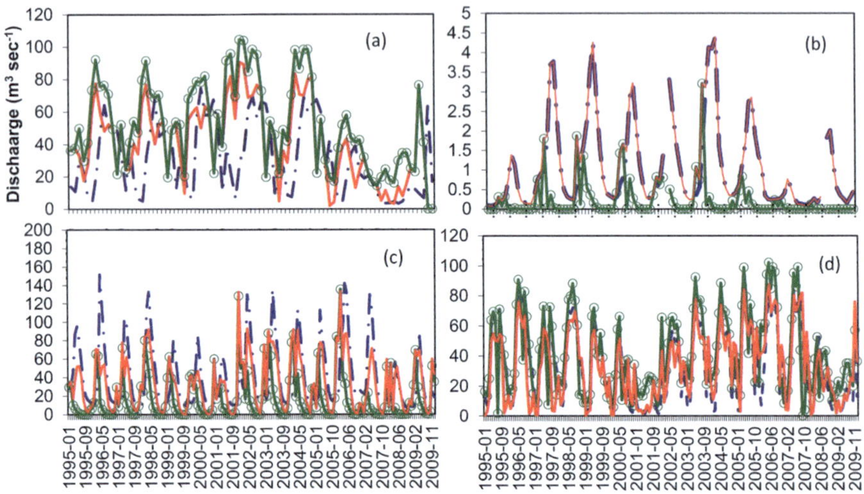

Fig. 14.8 Example of stations located at downstream of various natural and anthropogenic objects: Diziche station downstream of few check dams (**a**), Ghalenazer station downstream of Morghab spring (**b**), Ghaleshahrokh station downstream of Koohrang water transfer tunnels (**c**), and Polekale downstream of SadTanzimi dam (**d**). The *red solid line* and the green circled line show the swat simulated stream flow with and without simulating the effects of management or natural objects, respectively. The *blue dash line* shows measured stream flow data

(1) water diversion channels including traditional and modern irrigation networks, (2) reservoirs, dams (e.g., Zayandeh Rud dam), (3) springs, (4) extraction wells located at river bank, (5) inter-basin water transfers, and (7) geographic coordinates of some stations which were not properly reported and were located on a wrong tributary (Fig. 14.8).

A first run of the SWAT hydrological model, prior to calibration, revealed a considerable stream flow contribution from Shoor, Dastkan (Northern Tributaries), and Morghab rivers (see Fig. 14.3) into Zayandeh Rud main river (Fig. 14.9). This does not correspond to the actual condition. In actual condition, the Zayandeh Rud river does not receive water from these tributaries. This over estimation was mainly due to the water use data which was not available and therefore was not considered in the model. To increase model accuracy in representing the actual processes we ran several stakeholder meetings to understand the water management options and allocation scheme in theses tributary catchments. Therefore, we considered water allocation to various uses within these sub catchments in the model and also adjusted physical parameters to represent actual processes related to the rain water harvesting projects in the northern tributaries to allow the surface water to be infiltrated and recharged into the ground water (Fig. 14.10).

After identifying and properly accounting for management and all other natural and anthropogenic changes (Figs. 14.8 and 14.9), we calibrated hydrological model of the basin using the discharge data of 17 hydrometric stations with the regional approach as described in the previous section (Fig 14.10). It is important to mention

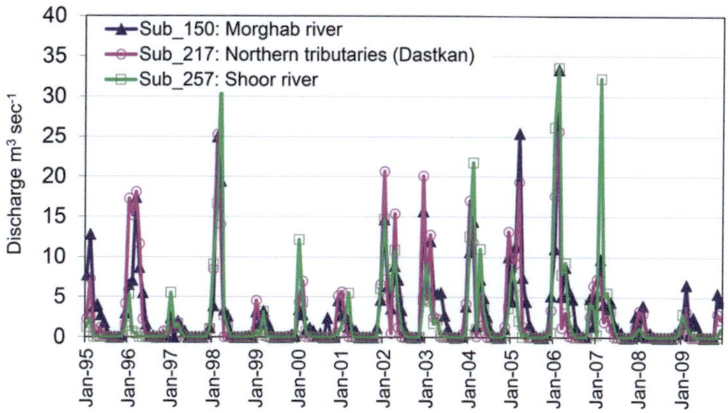

Fig. 14.9 Simulated stream flow contribution of the three main tributaries into Zayandeh Rud river prior to incorporating the water use and other management options in the model

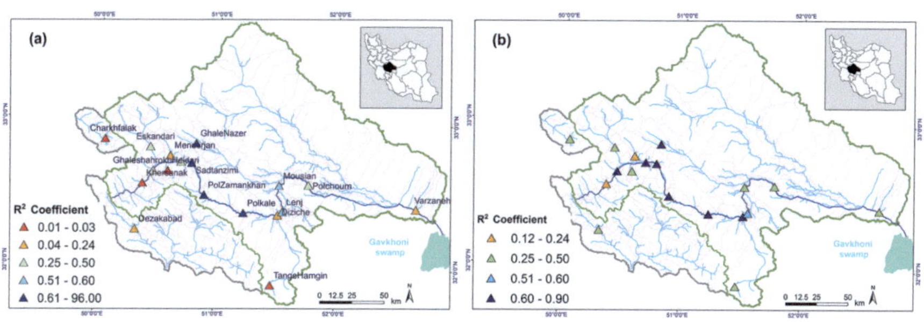

Fig. 14.10 Comparison of observed and simulated discharges using coefficient of determination (R^2) for 17 hydrometric stations: pre-calibration (**a**), post-calibration (**b**)

that an appropriate model setup incorporating the most important processes prevents over calibration of the physical models and increases model reliability in future scenario analysis (Faramarzi et al. 2015). Engagement of the stakeholders in various steps of the model development prevents this over calibration problem as it helps understanding the actual processes and increases our ability to setup a more representative model prior to calibration. Table 14.3 and Fig. 14.10 show our model performance for pre- and post-calibration steps. Overall, the basin wide bR^2 was improved from the average of 0.34 to about 0.45. However, performance gain was not identical for all stations and all sub catchments. In upstream region (see Fig. 14.3), the average bR^2 improved from 0.27 to about 0.44 while it improved from 0.39 to about 0.46 in downstream catchment. In upstream highland areas where snow is dominant in the winter season, snow hydrology and related processes were reasonably simulated with the best available data (see Fig. 14.7). However, the simulation results were not consistent for all stations in upstream catchment.

Table 14.3 Model performance of the study area during the pre- and post-calibration steps

		Pre-calibration			Post-calibration		
Sub-catchments	Station name	R^2	NS	bR^2	R^2	NS	bR^2
Upstream dam	Eskandari	0.22	−1.63	0.18	0.44	0.39	0.23
	Ghaleshahrokh	0.49	0.44	0.29	0.85	0.83	0.80
	Heidari	0.11	−42.00	0.09	0.29	−12.00	0.20
	Khersanak	0.13	−0.21	0.08	0.20	0.02	0.11
	Menderjan	0.22	−28.83	0.08	0.12	−4.35	0.09
	Sadtanzimi	0.87	0.85	0.84	0.87	0.83	0.86
	Zayand. Dam inflow	0.44	0.24	0.32	0.86	0.84	0.82
	Average	**0.35**	**−10.16**	**0.27**	**0.52**	**−1.92**	**0.44**
Downstream dam	Dizicheh	0.62	0.55	0.49	0.81	0.76	0.68
	Lenj	0.48	0.03	0.42	0.60	0.27	0.44
	Mousian	0.44	−0.08	0.40	0.49	0.14	0.48
	Polechoum	0.26	−1.61	0.25	0.30	−0.60	0.15
	Polekale	0.71	0.68	0.59	0.75	0.70	0.60
	Polezamankhan	0.82	0.76	0.76	0.82	0.76	0.76
	Tangehamgin	0.23	0.17	0.15	0.36	0.26	0.25
	Varzaneh	0.09	-89.70	0.03	0.33	0.29	0.31
	Average	**0.46**	**−11.15**	**0.39**	**0.56**	**0.32**	**0.46**
Karoon-Dez	Charkhfalak	0.32	0.31	0.19	0.47	0.39	0.23
	Dezakabad	0.29	0.19	0.18	0.42	0.39	0.23
	Ghalenazer	0.78	0.52	0.69	0.90	0.66	0.86

Inadequate availability of the temperature data in Ghaleshahrokh sub-catchment (southern tributary in upstream dam) caused poorer performance (e.g. Khersanak station, Fig. 14.11a) than northern tributaries (e.g. Eskandari station, Fig. 14.11b) where more climate data were employed in the model from the nearest stations. The quantity of temperature data for upstream dam, especially in Ghalehshahrokh basin is poor and simulation of snow fall and snow melt for all modeled sub-basins located at Ghalehshahrokh basin is based on the single temperature station in Ghaleshahrokh region. In downstream stations the calibration performance was highly depended on the quality and quantity of water use and water diversion data. As shown in Fig. 14.11c, d the calibration results after Polekale station are not as desirable as those in upstream stations (see Table 14.3).

14.8 Water Balance

To further verify the model results we plotted Fig. 14.12. We aggregated our sub-basin based and monthly predictions to Zayandeh Rud basin only and compared water balance components with the results of the study conducted by Water

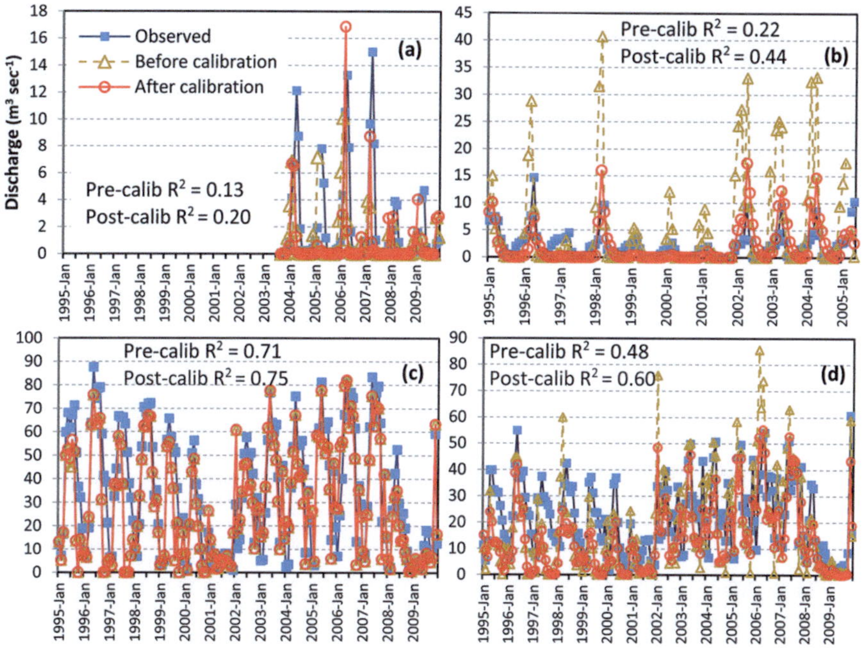

Fig. 14.11 Comparison of the observed (*blue squared blue line*) and simulated (*red circled line* for post calibration and *triangle yellow line* for pre-calibration steps) river discharges for the 1995–2009 period for selected stations: Khersanak (**a**), Eskandar (**b**), Polekle (**c**), and Dizicheh (**d**) stations

and Sustainable Development (WSD) in Iran for the last 30 years. As shown in Fig. 14.12a, our results are comparable with the study by (WSD) in Isfahan.

Upstream dam is the major source of water supply for the downstream region. To verify our results in upstream dam we plotted Fig. 14.12b–d to account for the volume of water entered into the Chadegan Reservoir (Zayandeh Rud Dam). As shown there is a good match between observed and simulated stream flow (Fig. 14.12b); and the monthly and yearly variation of water entering into the reservoir (Fig. 14.12c, d) were reasonably simulated. Overall, our simulation resulted in 1300 ± 20 MCM of water inflow to the reservoir over 1995–2009 whereas the observed data produced 1349 MCM of water entering into Zayandeh Rud Dam over this period of time.

To meet the ultimate goal of this project, the naturalized stream flow daily data was required from SWAT to feed the MIKE Basin (MB) model for the management and allocation purposes. To generate the natural stream flow using SWAT, we used our calibrated and validated hydrologic model of the basin and excluded all management measures (i.e. water transfers, water diversion for agricultures, industries, drinking and municipalities, and dam operation) from the SWAT model and ran it on daily basis to simulate daily natural stream flow (DNS) for each SWAT sub-basin. The simulated DNS data in each SWAT sub-basin was spatially

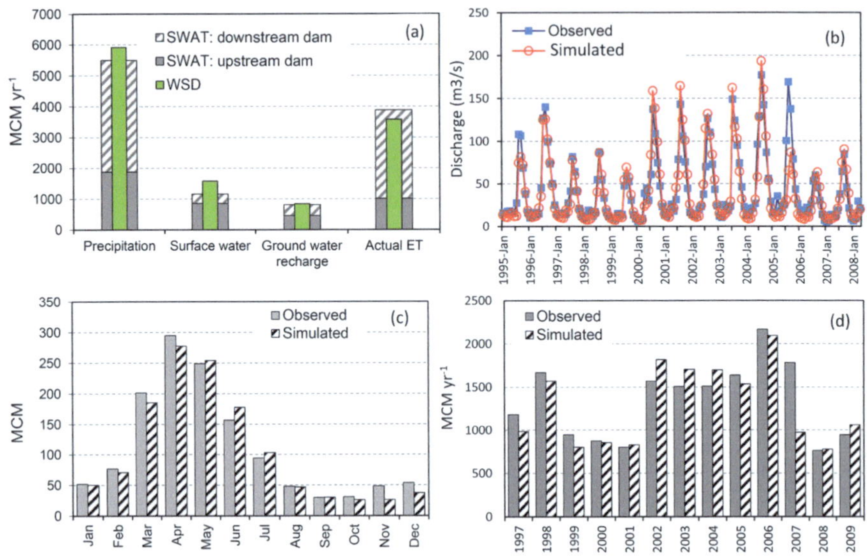

Fig. 14.12 Model verification through comparison of the simulated versus observed/reported data: water balance components (**a**), monthly stream flow entering the Zayandeh Rud Dam (**b**), long-term average (1995–2009) monthly volume of water entering Zayandeh Rud Dam (**c**), and total annual volume of water entering Zayandeh Rud Dam (**d**)

aggregated to account for the data of the MB-delineated catchments. Further, the simulated daily stream flows ($m^3\ sec^{-1}$) was converted to liter $km^{-2}\ day^{-1}$ for each MB catchment. Figure 14.13 shows the SWAT sub-basins and the MB catchments for our study area where the two models interact in this project.

14.9 Remarks and Recommendations

In large scale hydrological models, precision of the parameter estimation depends on the quality and quantity of the available input data. In this study the available data generally allowed obtaining satisfactory results, but inclusion of a larger number of climate stations, especially in upstream highland terrain, could have improved the quality of the predictions. An advanced research study is required for better quantification of the hydrological behavior in tributary river basin (i.e. Shoor, Morghab and Dastkan). Calibration of the SWAT hydrologic model against other water cycle components (e.g. soil moisture, ground water recharge) rather than stream flow will increase model reliability.

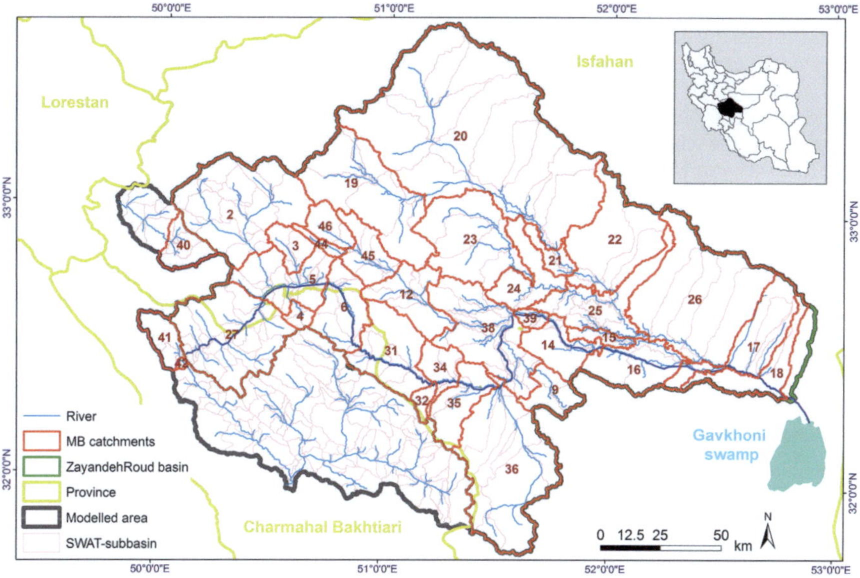

Fig. 14.13 Study area representing the 370 SWAT sub-basins and 33 MIKE BASIN catchments

References

Abbaspour KC (2011) User manual for SWAT-CUP: SWAT calibration and uncertainty analysis programs. Eawag: Swiss Federal Institute of Aquatic Science and Technology, Duebendorf, 103 pp

Arnold JG, Srinivasan R, Muttiah RS, Williams JR (1998) Large area hydrologic modeling and assessment Part 1: model development. J Am Water Resour Assoc 34:73–89. doi:10.1111/j.1752-1688.1998.tb05961.x

Bennett ND, Croke BFW, Guariso G, Guillaume JHA, Hamilton SH, Jakeman AJ, Marsili-Libelli S, Newham TH, Norton JP, Perrin C, Pierce SA, Robson B, Seppelt R, Voinov AA, Fath BD (2013) Characterizing performance of environmental models. Environ Model Softw 40:1–20. doi:10.1016/j.envsoft.2012.09.011

Doll P, Berkhoff K, Fohrer N, Gerten D, Hagemann S, Krol M (2008) Advances and visions in large-scale hydrological modelling: findings from 11th workshop on large-scale hydrological modelling. Adv Geosci 18:51–61. doi:10.5194/adgeo-18-51-2008

Faramarzi M, Abbaspour KC, Schulin R, Yang H (2009) Modelling blue and green water resources availability in Iran. Hydrol Process 23:486–501. doi:10.1002/hyp.7160

Faramarzi M, Yang H, Mousavi J, Schulin R, Binder CR, Abbaspour KC (2010) Analysis of intra-country virtual water trade strategy to alleviate water scarcity in Iran. Hydrol Earth Syst Sci 14:1417–1433. doi:10.5194/hess-14-1417-2010

Faramarzi M, Abbaspour KC, Vaghefi A, Farzaneh S, Zehnder M, Srinivasan AJB, Yang H (2013) Modeling impacts of climate change on freshwater availability in Africa. J Hydrol 480:85–101. doi:10.1016/j.jhydrol.2012.12.016

Faramarzi M, Srinivasan R, Iravani M, Bladon KD, Abbaspour KC, Zehnder AJB, Goss G (2015) Setting up a hydrological model of Alberta: Data discrimination procedure prior to calibration. Environ Model Softw 74:48–65. doi:10.1016/j.envsoft.2015.09.006

Hargreaves G, Samani ZA (1985) Reference crop evapotranspiration from temperature. Appl Eng Agric 1:96–99. doi:10.13031/2013.26773
Jarvis A, Reuter HI, Nelson A, Guevara E (2008) Hole-filled SRTM for the Globe Version 4. In: The CGIAR-CSI SRTM 90 m Database. http://srtm.csi.cgiar.org
Legesse D, Vallet-Coulomb D, Gasse F (2003) Hydrological response of a catchment to climate and land use changes in Tropical Africa: case study south central Ethiopia. J Hydrol 275:67–85. doi:10.1016/S0022-1694(03)00019-2
Neitsch SL, Arnold JG, Kiniry JR, Williams JR (2011) Soil and water assessment tool. Theoretical documentation: Version 2000. TWRITR-191. Texas Water Resources Institute, College Station, TX
Qiu LJ, Zheng FL, Yin RS (2012) SWAT-based runoff and sediment simulation in a small watershed, the loessial hilly-gullied region of China: capabilities and challenges. Int J Sed Res 27:226–234. doi:10.1016/S1001-6279(12)60030-4
Ritchie JT (1972) A model for predicting evaporation from a row crop with incomplete cover. Water Resour Res 8:1204–1213. doi:10.1029/WR008i005p01204
Schoul J, Abbaspour KC, Yang H, Srinivasan R (2008) Modelling blue and green water availability in Africa. Water Resour Res 44:1–18. doi:10.1029/2007WR006609
Tang FF, Xu HS, Xu ZX (2012) Model calibration and uncertainty analysis for runoff in the Chao river basin using sequential uncertainty fitting. Proc Environ Sci 13:1760–1770. doi:10.1016/j.proenv.2012.01.170

Chapter 15
Application of the FEFLOW Groundwater Model in the Zayandeh Rud Catchment

Sebastian Sklorz, Michael Kaltofen, and Bertram Monninkhoff

15.1 Introduction

The groundwater model was build up with the commercial software package FEFLOW (Diersch 2014), which calculates water flow, mass and heat transport in porous media. For a better understanding, the groundwater model will be named FEFLOW Model in the remainder of this paper. The aim of the FEFLOW model was to identify the volumetric amount of seepage water in relation to the groundwater levels nearby the Zayandeh Rud River. The area under investigation has an extension of 22,868 km^2 and covers nine sub-catchments (codes: 4201, 4202, 4203, 4205, 4206, 4207, 4208, 4209 and 4217). Not the entire area under investigation is part of the FEFLOW model; rather, the main focus is the porous aquifer system nearby the Zayandeh Rud River. Therefore, the area for the FEFLOW model was restricted to the aquifer extension. An overview of the aquifer extension, the areas of the FEFLOW, SWAT and the MIKE BASIN model is presented in Fig. 15.1. The FEFLOW model covers an area of 10,446 km^2

15.2 Model Setup

The boundary to the northern part is defined by the aquifer extension in sub-catchment code 4205. In the north-western part, a watershed between sub-catchment codes 4207 and 4214 defines the boundary of the FEFLOW model. The aquifer in sub-catchment code 4204 is already partly dry. Along with the narrow bottleneck connection to the aquifer in the neighboured sub-catchment

S. Sklorz (✉) • M. Kaltofen • B. Monninkhoff
DHI-WASY GmbH, Berlin, Germany
e-mail: ssk@dhigroup.com

S. Mohajeri, L. Horlemann (eds.), *Reviving the Dying Giant*,
DOI 10.1007/978-3-319-54922-4_15

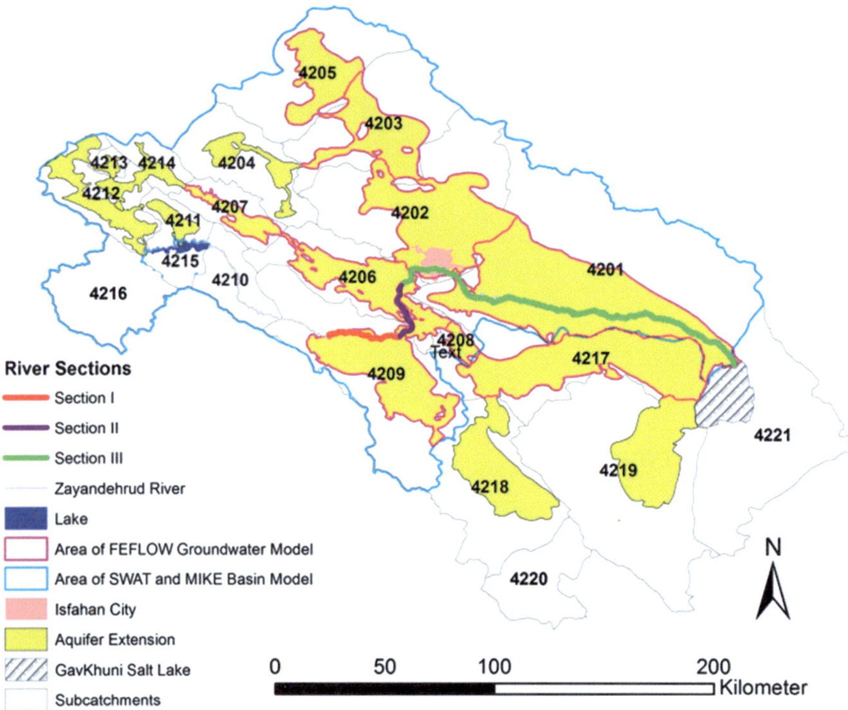

Fig. 15.1 Overview of the extension of the three models, the extent of the aquifer and the division of flow sections for the exchange between the MIKE BASIN and the FEFLOW models (Mohajeri et al. 2016)

code 4203, this led to the assumption that the sub-catchment code 4204 has no influence on the water exchange between Zayandeh Rud River and the nearby aquifer. In the south, the aquifer extensions of sub-catchments 4217 and 4209 were used as boundaries for the FEFLOW model. For the sub-catchments 4218, 4219, 4220 and 4221, the available data was insufficient for a reliable implementation within the FEFLOW model or the water exchange was considered as not relevant for the interaction between the aquifer and the Zayandeh Rud River (DHI-WASY 2014).

Little is known about the geological layering of the alluvium aquifer in the study area, given that no interpreted geological cross sections were available. Only the existence of one approximately 10-metre-thick clay layer in sub-catchment code 4201 is widely accepted by the Esfahan Regional Water Board. This clay layer separates the upper unconfined from the lower aquifer, for which it is assumed to have confined conditions. The top, bottom and extension of the clay layer were defined by data from the Iranian partners. The sedimentary lithofacies show imbedded coarse gravel/sandy deposits of proximal to muddy, fine-grained deposits of distal areas of Quaternary dry land alluvial megafans (Arzani 2012). Little usable data for a stratified and layered geological model was available. Excluding the clay

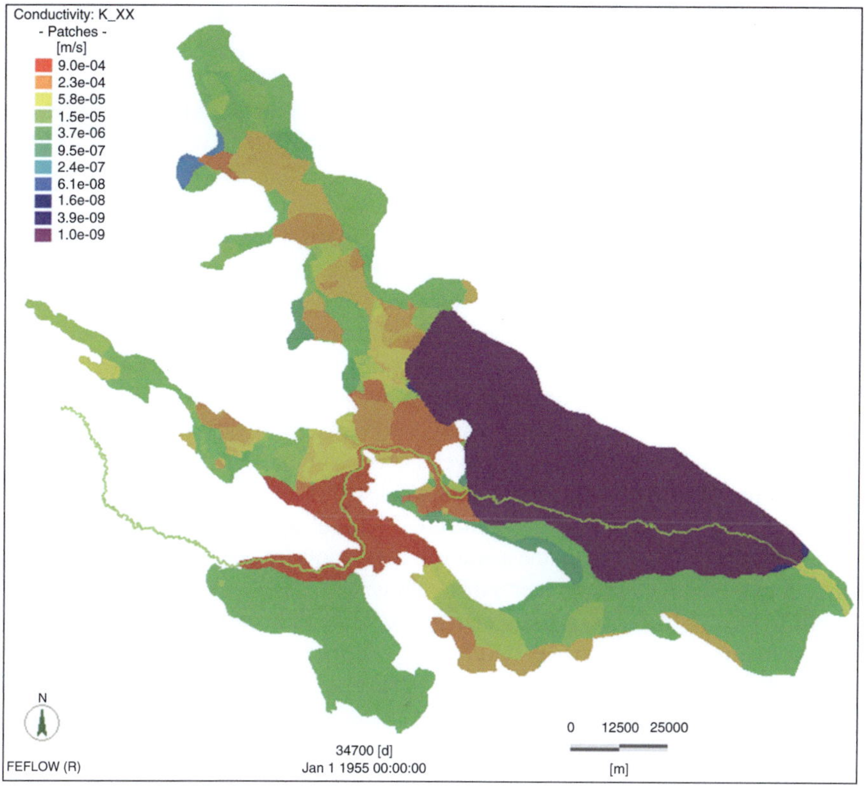

Fig. 15.2 Extension of the clay layer together with the distribution of hydraulic conductivity (DHI-WASY 2014)

layer, the rest of the model was assumed to comprise a vertically homogeneous geology. The extension of the clay layer is illustrated in Fig. 15.2.

In the study area, three major groundwater extraction methods are applied, namely springs, wells and qanats. Springs are natural sites where groundwater comes out of the sub-surface and becomes surface water. A comparison between the location of springs and geological fault structures indicates a clear correlation between these two. Wells are all kinds of man-made structures where groundwater is lifted by electric pumps or other mechanical lifting methods. Qanats are traditional artificial systems for extracting groundwater at topographic slopes by horizontal shafts burrowed into the groundwater. A detailed description of the functionality of qanats is provided by Wulff (1968). In the FEFLOW model, the extraction of more than 1300 springs, almost 1000 qanats and approximately 38,500 wells was considered. While the extraction of springs and qanats mostly depends on natural processes, the extraction of wells can be managed by the operators. The average groundwater extraction by wells in the model for the period from 1999 to 2012 was 2482 million m^3/year. Considering that the average qanat and spring extraction equals 214 million m^3/year, the total groundwater extraction in the model area is approximately 2696 million m^3/year.

The groundwater extraction was established in the FEFLOW model by using the two parameters "In/Outflow on Top/ Bottom" and "Source/Sink". Two different parameters were used to separate between deep and shallow wells. Deep wells have a depth of more than 50 m, while shallow wells extract at depths below 50 m. The Esfahan Regional Water Board (ERWB) provided yearly values of groundwater extraction for the overall river basin (Gavkhuni Basin) for the period from 2007 to 2010 (Iranian calendar year 1385 to 1388). From this data, a percentage-based depth coefficient ("*Depth_coef*") was calculated by considering the average extraction rates of all 3 years. The coefficient defines the extraction per sub-catchments in proportions of deep and shallow extraction. Yearly data of groundwater extractions was only available on a sub-catchment scale. However, for the FEFOW model, the spatial distribution of groundwater extraction is an important issue. Therefore, the average extraction capacities of all wells were summarised on a 1000*1000 m raster. In the following step, the percentage distribution of the average extraction capacity per sub-catchment was calculated. The resulting coefficient ("*Perc_coef*") represents the proportion of the summed extraction in one square kilometre in relation to the total extraction per sub-catchment.

FEFLOW calculates on a triangular mesh, while the coefficient *Perc_coef* was calculated on a square basis. Therefore, it was necessary to implement a coefficient that takes different areas of the square and the selected triangles into account. This coefficient was named "*Area_coef*" and its calculation is shown by means of an example in Fig. 15.3.

As a result, three dimensionless coefficients were generated. By multiplication with the observed extraction rate (for every time step in the available time series), a correct representation of the groundwater extraction rate for the FEFLOW model could be calculated. The following equation was used:

$$GW_{ext} = Perc_{coef} * Area_{Coef} * Depth_{Coef} * \text{time series}$$

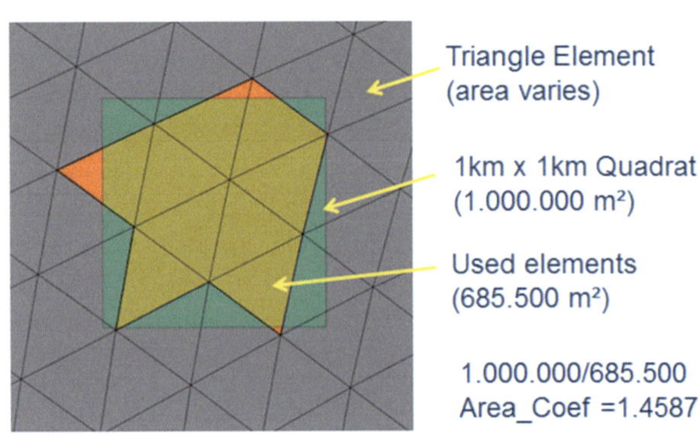

Fig. 15.3 Example for the calculation of the coefficient Area_coef (DHI-WASY 2014)

The groundwater extraction and the so-called groundwater return flow are closely connected processes. The groundwater return flow denotes a process where irrigated water percolates through the unsaturated zone and recharges the aquifer. This artificial recharge depends on several processes and parameters. Respective data to gain more insights into these processes was not available. Therefore, a simplified approach for the groundwater return flow was implemented. The groundwater return flow was assumed as product of the groundwater extraction (groundwater extraction by wells, springs and qanats) and a spatial variable coefficient, which represents the amount of surface water being used for agriculture irrigation. The implementation of the groundwater return flow in the FEFLOW model was performed by using the parameter "Source/Sink" in the first layer. The groundwater return flow was automatically linked to the groundwater extraction. Groundwater extraction and groundwater return flow only take place as long as the aquifer is not dry. In case no water is available, the groundwater extraction and the groundwater return flow are automatically switched off. The mean artificial groundwater recharge by irrigation implemented in the FEFLOW model amounts to 161 mm/a. In arid environments, in which irrigation is predominantly applied as flood irrigation, Global Water Partnership (2012) reported values between 70 and 300 mm/a. Based upon data from Consulting Engineers Water and Sustainable Development (2010), the artificial recharge in the area under investigation amounts to approximately 179 mm/a, which fits well to the amount implemented in the FEFLOW model.

The natural groundwater recharge was calculated by a separate SWAT model (Faramarzi and Besalatpour 2014). An average natural groundwater recharge of 0–2 mm/a (around 1% of the yearly precipitation) is common for arid environments. In semi-arid to arid environments, groundwater recharge rates of 8 mm/a are suggested by Gräbe (2012). Nikouei et al. (2012) analysed several sub-catchments in the Zayandeh Rud basin and differentiated between an average groundwater recharge of 13 and 2 mm/a (the latter represents groundwater recharge during drought periods). Due to the low percolation to the sub-surface and the high surface water run-off at hard rock mountains, Gieske and Miranzadeh (2000) described a relatively high groundwater recharge near the mountains. For the FEFLOW model, two natural types of groundwater recharge were defined. Recharge occurs on flat land when precipitation percolates down to the sub-surface and finally reaches the groundwater ("*GWR_A*"). At the borders of the aquifer, the alluvial aquifer is fed by additional mountain front recharge ("*GWR_M*"). A schematic model concept and vertical discretisation of the 3D-FEFLOW model is shown in Fig. 15.4.

The elevation of slice 1 was defined by the available digital elevation model. The top and bottom of the clay layer (slice 4 and slice 5) were defined by data from the Esfahan Regional Water Board. The bottom of the FEFLOW model was defined by the aquifer base, which was also provided by the Iranian partners.

The Zayandeh Rud River was defined with a vertical 3rd kind (Cauchy) boundary condition on slice 1 and slice 2. The exchange occurs at both sites of the vertical area represented by these boundary conditions. As the thickness of the first layer equals approximately 25 m, a good representation of the average river infiltration width of 50 m could be guaranteed with this method. Measurements of surface

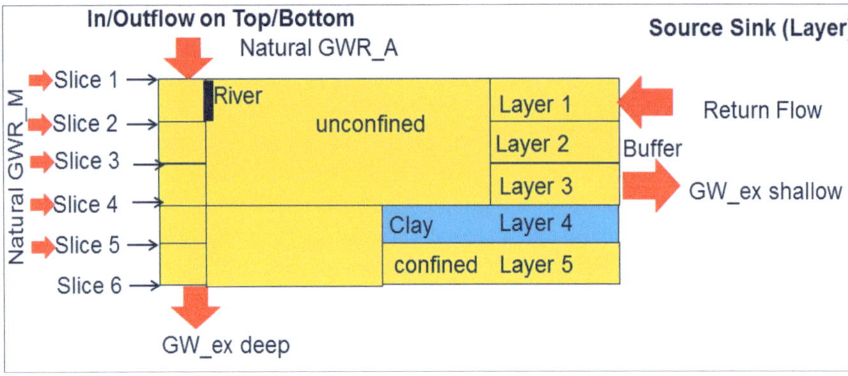

Fig. 15.4 Schematic model concept and vertical discretisation of the 3D-FEFLOW model (DHI-WASY 2014)

water heads were available at four stations, namely Pole-Kalleh, Dizicheh, Lenj and Moursian. While the four stations cover around 60 km of the total river length, for the majority of the river length (150 km) no surface water measurements were available. For these areas, surface water levels were derived by means of topography data as well as free satellite images.

15.3 Calibration

The FEFLOW model was build up as an unsaturated model because even during the 1995–2009 period several parts of the aquifer became dry. For the unsaturated flow, porosity data from Consulting Engineers Water and Sustainable Development (2010) and Safavi and Bahreini (2009) was used. The model was build up as a transient model with constant boundary conditions. The model was calibrated for the 1995–2009 period mainly by manual variation of the parameters hydraulic conductivity and the return flow coefficient. From the total number of 311 observation wells, 55 are in a range of 5 km around the river. Accordingly, these 55 observation wells were classified as relevant and considered during the calibration process. The scatter plot for the relevant 55 observation wells is presented in Fig. 15.5, showing that the entire GWM has an acceptable quality, although observed and calculated heads can differ by several meters in some parts of the model. The root-mean-square deviation for these wells equals 10.6 m. To facilitate the comparison between datasets or models with different scales, usually this RMS value is normalised with respect to the range of observed values within the model domain. In this case, the groundwater levels vary between 1470 and 1860 m, resulting in a normalised NRMS value of less than 3%, which is good for such a large-scale model (DVGW 2014).

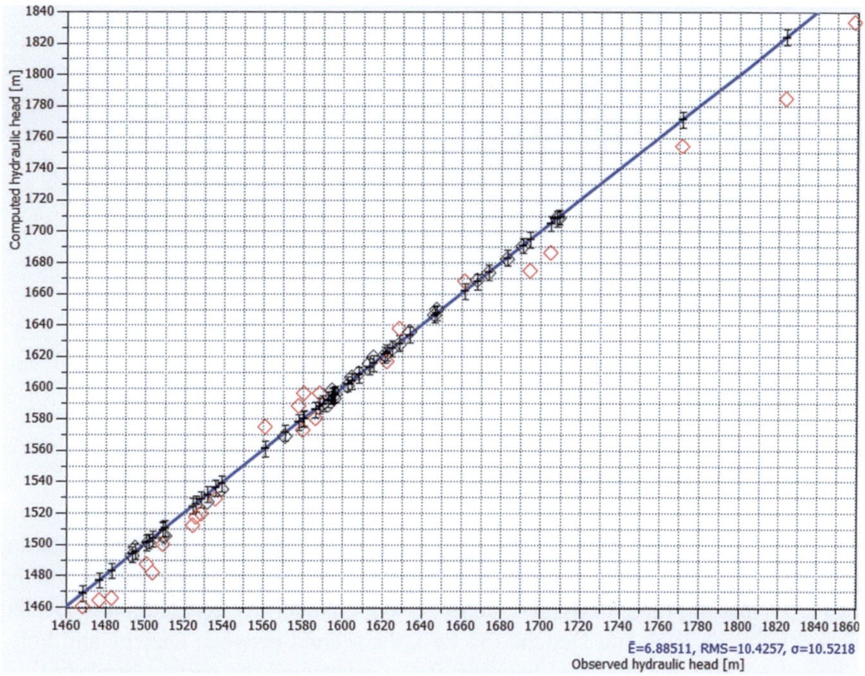

Fig. 15.5 Scatter plot the 55 relevant observation wells

To demonstrate the quality of the transient model calibration with respect to the change of groundwater levels within the simulation period, a cross-section view through sub-catchment 4206 (Najafabad) is shown in Fig. 15.6.

The cross-section view shows a good correlation between the observed and calculated heads at the selected section for the end of 2009. Furthermore, it shows the strong drawdown of more than 60 metres in the centre in only 15 years. The river leakage causes a strong inflow into the FEFLOW model and causes a significant smaller drawdown in the area near the river.

15.4 Groundwater and Surface Water Interaction

Regarding the groundwater resources, the extensive over-use of groundwater can be clearly illustrated by the groundwater levels, which have been decreased locally by more than 50 m during the last 15 years. Downstream of Pol Kale, the Zayandeh Rud River is losing water to the groundwater, in a process commonly known as river leakage.

As long as the river can be sufficiently recharged by releases from the main dam to avoid dryness in the river bed and as long as the groundwater levels are not

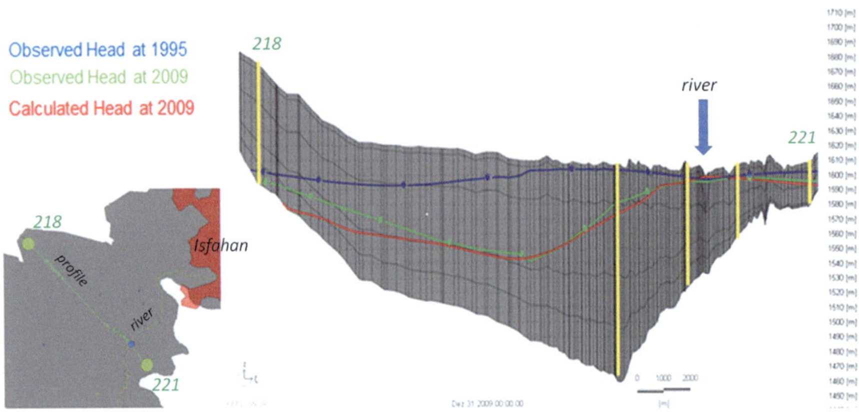

Fig. 15.6 Cross-section views through sub-catchment 4206 (Najafabad) with observed heads at 1995 and 2009 and calculated heads in 2009 (Mohajeri et al. 2016)

completely decoupled from the river, the leakage rate increases linear with falling groundwater levels. For the exchange between the Water Management Tool and the FEFLOW model, three exchange sections were defined: the first section is located between Pole-Kalleh and Diziche (33 km), the second between Diziche and Pole-Choum (64 km) and the third between Pole-Choum and the eastern border of the FEFLOW model (116 km). The sections are illustrated in Fig. 15.7 and the calculated surface water leakage for the three sections is presented in Fig. 15.8.

Because no seasonal dry parts of the river were considered in the FEFLOW model, the exchange rates increased with decreasing groundwater levels. During the first three years, the surface water leakage strongly increases to reach saturated conditions in the relevant parts of the model. After this time, the model becomes stable and calculates a nearly linear increase for the surface water leakage. The river leakage in Sect. 15.1 did not show major changes from 1998 to December 2009, which can be explained with relatively constant groundwater levels in the adjacent aquifer during the simulation time in that section. Section 15.2 passes though sub-catchment code 4206 (Najafabad), where intensive agriculture activities with high groundwater extraction rates cause strong groundwater level decreases. The ongoing decrease of the groundwater level causes an increase in surface water leakage, whereby the same process took place in Sect. 15.3.

For Sect. 15.1, the exchange rate was quantified with approximately 32 million m^3/a. For Sects. 15.2 and 15.3, the exchange rates were quantified with 185 and 268 million m^3/a, respectively, as a weighted average over the simulation time.

The FEFLOW model calculates a total net exchange between surface water and groundwater of 484 million m^3/a. For the FEFLOW model, this positive net exchange results in water gains and thus this number represents an overall surface water flow to groundwater. The calculated net exchange is strongly comparable with the data from Consulting Engineers Water and Sustainable Development (2010), which presented a surface water to groundwater exchange of 525 million m^3/a. Safavi et al. (2015) presented yearly data of groundwater recharge from the

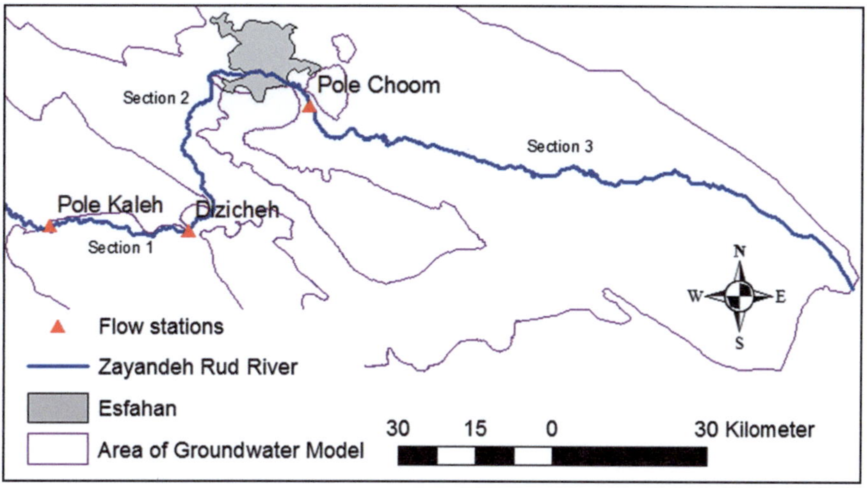

Fig. 15.7 River sections for evaluating surface water and groundwater exchange (DHI-WASY 2014)

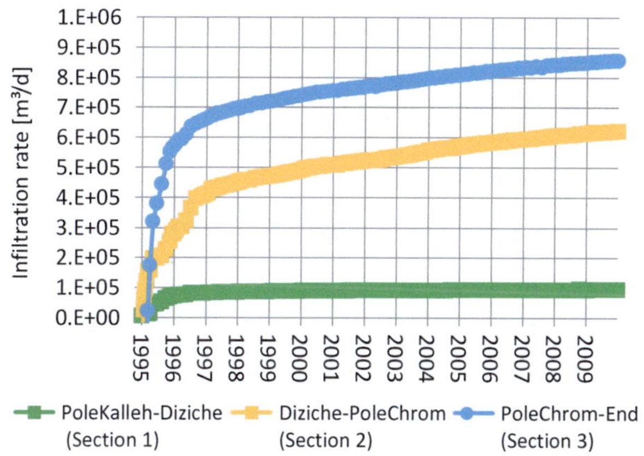

Fig. 15.8 Surface water leakage at different sections (DHI-WASY 2014)

Zayandeh Rud River to the adjacent aquifers of approximately 400 million m^3/a for the period from 1991 to 2011. Based upon the river length of 213 km and the average river width of 50 m, the calculated exchange rate is 0.13 m^3/m^2*d.

The general principle of surface water-groundwater interaction is shown in Fig. 15.9. In many systems, natural conditions involve groundwater levels being higher than the surface water level and thus the surface water discharge being increased by groundwater inflow. Along the Zayandeh Rud River, these conditions are currently hardly present. As a result, a constant leakage from surface water to the aquifers is observed. Without a significant decrease of groundwater extraction,

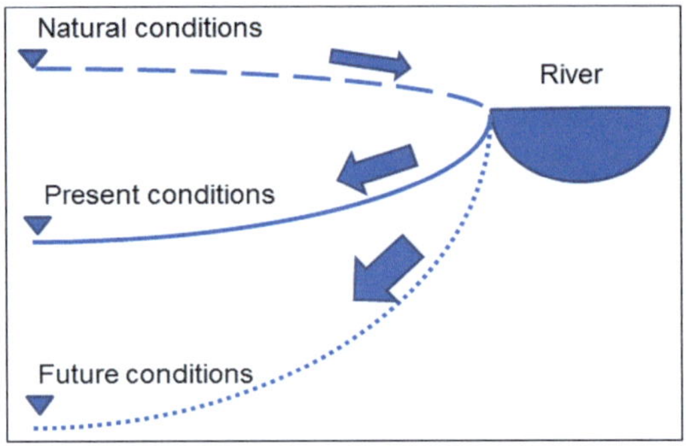

Fig. 15.9 Schematic illustration of the surface water—groundwater interaction

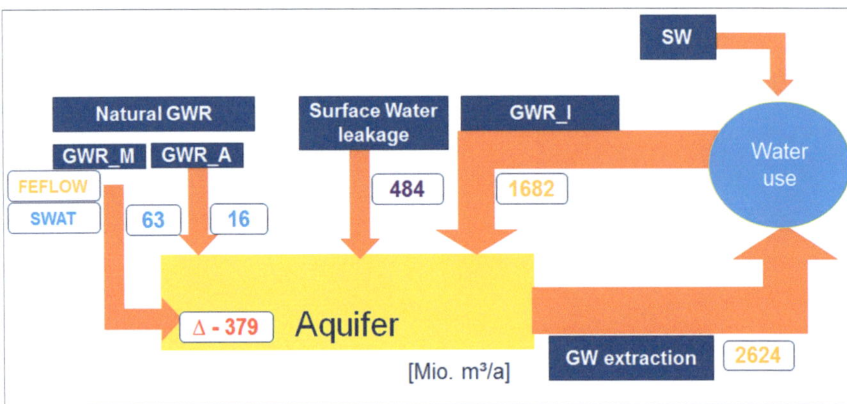

Fig. 15.10 Budget of the FEFLOW model for the 1995–2009 period (DHI-WASY 2014)

the decrease of the groundwater level will continue, which also results in an increasing surface water leakage. The problem of partly and seasonal dry river beds along the Zayandeh Rud River is thus likely to increase as long as no adequate counteracting measures are found.

15.5 Model Budget

To calculate the model budget, all in- and outflows of the model were considered, whereby the budget for the FEFLOW model is illustrated in Fig. 15.10. In this balance, the SWAT results represent those areas that cause an inflow into the FEFLOW model, separated into two components representing the inflows at the plane (GWR_A) and the mountain (GWR_M) areas. Overall, the natural groundwater recharge for the FEFLOW model is around 79 million m^3/a, while the total groundwater extraction in the FEFLOW model over the model period equals 2624 million m^3/a. This amount marginally differs from the input extraction rate (2696 million m^3/a), showing that at some parts of the model the aquifer became dry and pumping was automatically stopped. Based upon this groundwater extraction, the return flow (groundwater recharge by irrigation, GWR_I) provided the major inflow to the FEFLOW model with 1682 million m^3/a. At this point, it has to be noted that there is significant use of surface water for irrigation, which cannot be explicitly quantified in the actual groundwater budget. The main modelling output for this stage of the overall project–namely the inflow from surface water to groundwater (surface water leakage)–is calculated at 484 million m^3/a.

References

Arzani N (2012) Groundwater recharges in Dryland alluvial megafans: geomorphology versus lithofacies controls: examples from Central Iran. Hydrogeology of Arid Environments, Stuttgart

Consulting Engineers Water and Sustainable Development (2010) Part V: integrate and water balance studies. Volume III: analysis of data and water balance Studies updating the Atlantic Basin Water Resources Study areas Gavkhooni, Iran

DHI-WASY (2014) Integrated water resource Management in Isfahan, Iran: groundwater model for the Zayandeh Rud catchment. DHI-WASY GmbH, Berlin. (unpublished)

Diersch HJ (2014) FEFLOW–finite element modeling of flow, mass and heat transport in porous and fractured media. Springer, Heidelberg/Berlin. doi:10.1007/978-3-642-38739-5

DVGW (2014) Technische regel—Arbeitsblatt W 107 (A), Aufbau und Anwendung numerischer Grundwassermodelle in Wassergewinnungsgebieten, DVGW Deutscher Verein des Gas- und Wasserfaches e.V. Bonn, Entwurf August 2014

Faramarzi M, Besalatpour AA (2014) Hydrological modelling of Zayandeh Rud river basin. Isfahan University of Technology–Department of Natural Resources, Isfahan

Gieske A, Miranzadeh M (2000) Groundwater resource modelling of the Lenjanat aquifer system. IAERI-IWMI Research Reports No. 15

Global Water Partnership (2012) Groundwater Resources and Irrigated Agriculture—making a beneficial relation more sustainable, www.gwptoolbox.org

Gräbe A et al (2012) Development of a 3D groundwater flow model in semi-arid to arid regions: the western drainage basin of the Dead Sea. Geophysical research abstracts, vol 14, European Geosciences Union (EGU), p EGU2012-8086. EGU General Assembly 2012, Vienna, Austria, 22–27 April 2012

Mohajeri S, Horlemann L, Sklorz S, Kaltofen M, Ghanavizchian S, Nunez von Voigt T (2016) Integrated water resource management in Isfahan: the Zayandeh Rud catchment. In: Borchardt D, Bogardi J, Ibisch R (eds) Integrated water resources management: concept, research and implementation. Springer, Basel. doi:10.1007/978-3-319-25071-7_23

Nikouei A, Zibaei M, Ward FA (2012) Incentives to adopt irrigation water saving measures from wetlands preservation: an integrated basin scale analysis. J Hydrol 464–465:216–232. doi:10.1016/j.jhydrol.2012.07.013

Safavi H, Bahreini G (2009) Conjunctive simulation of surface water and ground water resources under uncertainty. Iran J Sci Technol, Trans B, Eng 33(B1):79–94

Safavi H, Golmohammadi M, Sandoval Solis S (2015) Expert knowledge based modelling for integrated water resources planning and management in the Zayandehrud River basin. J Hydrol 528:773–789. doi:10.1016/j.jhydrol.2015.07.014

Wulff HE (1968) The qanats of Iran. Sci Am 218(4):94–105

Chapter 16
Application of MIKE Basin in the Zayandeh Rud Catchment

Michael Kaltofen, Fabian Müller, and Anna Zabel

16.1 Introduction

MIKE Basin is a decision support tool for integrated water resources analysis, planning and management of river basins (DHI 2012). The calculation of water availability in the MIKE Basin model is based upon a continuously in time performed balance of water yield and water use as well as water management. The balancing starts from natural run-off contributing from hydrological catchments and totalised from head basins down to outflows of the study area. This procedure results in natural stream flows at points of interest at the Zayandeh Rud River. Subsequently, water extractions and water management impacts such as dam regulations, water diversion and transfers as well as discharges i.e. from waste water treatment plants are applied, including percolation losses from the Zayandeh Rud River.

MIKE Basin has the feature of customising the balancing process through scripting, thus providing the opportunity to change almost all input data to adopt the balancing to special rules or water budget behaviour, such as the interaction of surface and groundwater. This feature was strongly used owing to the complexity of water use and management processes in the Zayandeh Rud basin. This applies in particular to dam management and water supply rules.

MIKE Basin Zayandeh Rud was proven plausible regarding data concerning water use and transfers as well as dam management and water supply rules. For this purpose, input data was made available for the period from 1379 (2000) to 1387 (2008).

M. Kaltofen (✉) • F. Müller • A. Zabel
DHI WASY GmbH, Berlin, Germany
e-mail: kal@dhigroup.com

S. Mohajeri, L. Horlemann (eds.), *Reviving the Dying Giant*,
DOI 10.1007/978-3-319-54922-4_16

16.2 Water Demand Modelling for Agriculture

Since the agricultural sector uses the majority of water resources in this region, a detailed water demand calculation for the most relevant species has been conducted in this study. In MIKE Basin, the general irrigation approach was selected based upon the FAO 56 guideline (Allen et al. 1998). An overview of the irrigation module and parameters is provided in Fig. 16.1.

Within the Zayandeh Rud catchment, 15 relevant irrigation networks have been identified (see Fig. 16.2).

An irrigation node in MIKE Basin represents an irrigation area comprising one or more irrigated fields that draw water from the same source(s). Most of the irrigation networks are represented by two separated irrigation nodes in the MIKE Basin model, one for the crops and another for the orchards, leading to a total of 25 MIKE Basin irrigation nodes.

Hence, the main parameters of the irrigation module include:

- the assignment of the crops and orchards to the irrigation systems;
- the definition and assignment of the irrigation methods;
- the generalisation of the crop and orchards calendar;
- the soil characteristics for each crop sequence;
- the climate and reference ET model selection;
- the deficit distribution method; and
- the surface water and groundwater supply.

An assessment of the crops and orchards was conducted for the Zayandeh Rud catchment and a selection was made containing species covering almost all water demand by irrigation. At the same time, the exclusion of these crops leads to a faster model runtime and an improvement of the overall data handling in MIKE Basin. Furthermore, the cultures apples, stone fruits and grainy fruits were merged to the category of "fruits". The resulting total cultivated area per network is shown in Table 16.1, which also indicates crops and orchards being separated into two irrigation nodes.

The calculation of irrigation demand is based upon the calculation of crop evapotranspiration and soil evaporation. Two models are available in MIKE Basin and both have been used in this setup:

- the "Dual crop coefficient model (FAO 56)" for all crops (except for rice); and
- the "Rice crop model" for the rice crops.

The dual crop model separates the transpiration and soil evaporation calculation. The required crop characteristics were chosen according to specific conditions and their variation in the Zayandeh Rud basin. In particular, the crop calendar was adjusted for use in MIKE Basin. It reflects sowing dates and the total duration of all development stages. In the Zayandeh Rud basin, the sowing and harvesting dates vary among the irrigation networks due to climate. To generalise these variations, an analysis of variations of all crop parameters for every crop was conducted. As an example, the method and results of generalisation are shown for wheat (see Fig. 16.3).

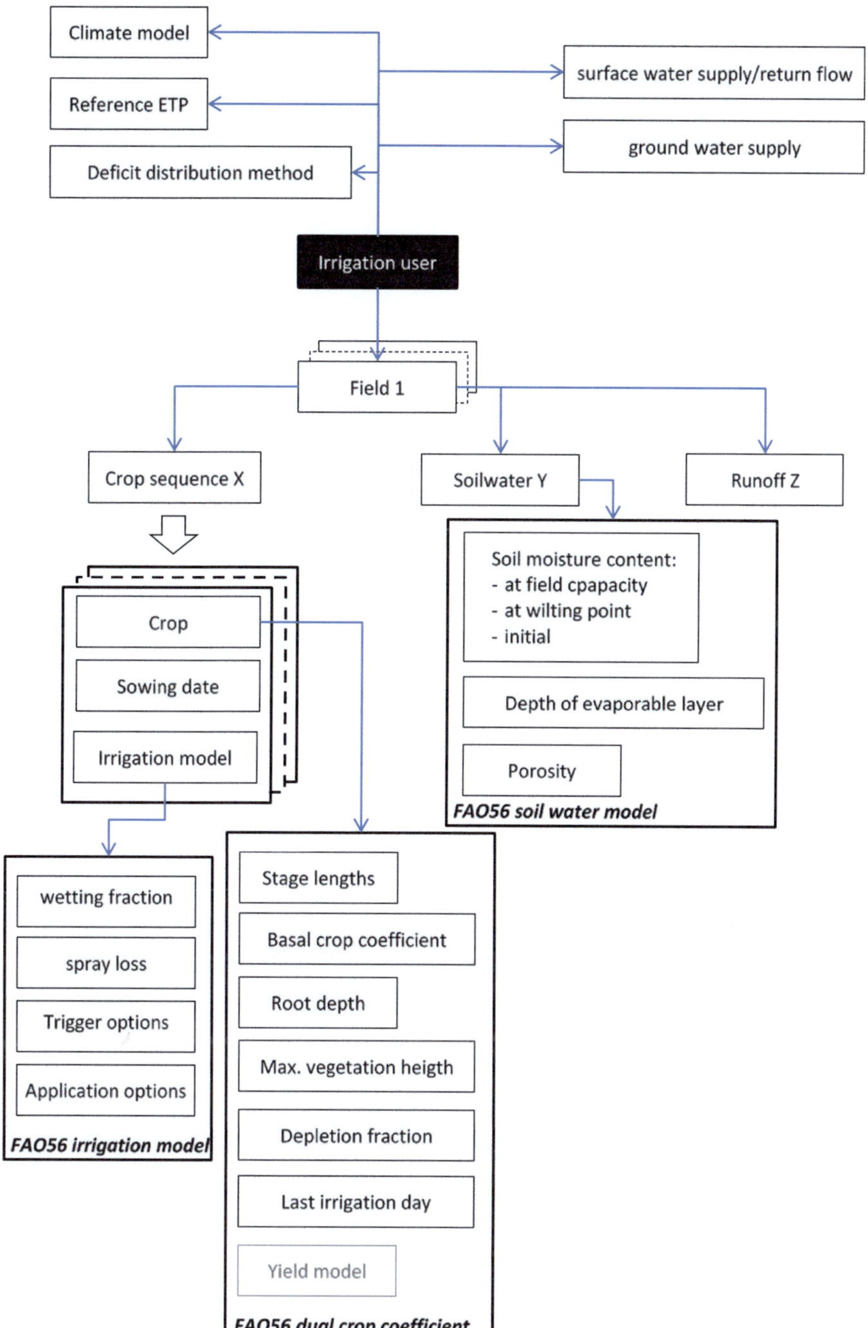

Fig. 16.1 Parameters of the irrigation module in MIKE Basin (based on FAO 56)

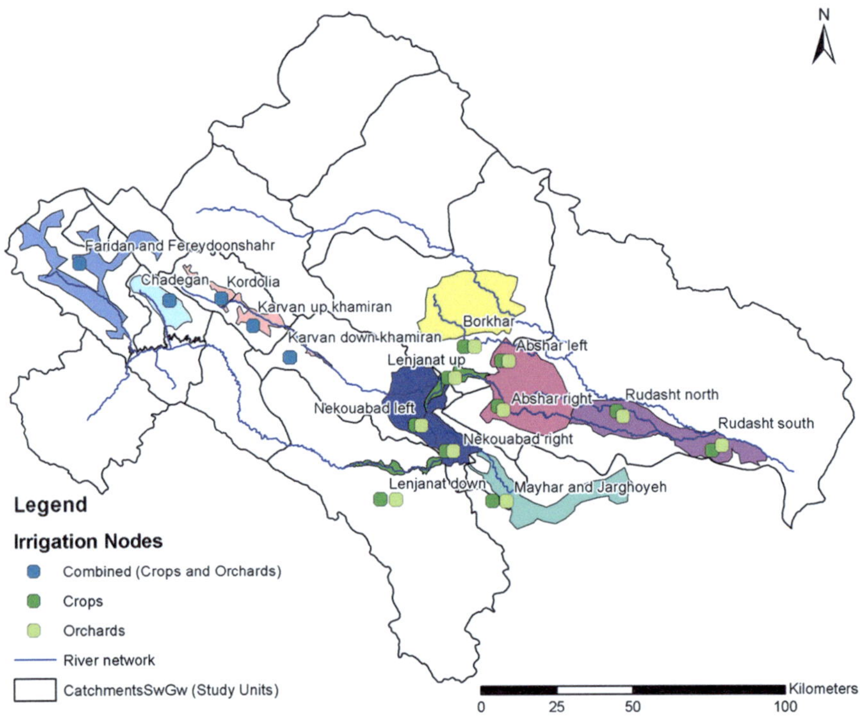

Fig. 16.2 Irrigation networks within the Zayandeh Rud catchments represented in the MIKE Basin model

As irrigation methods within the irrigation areas of the Zayandeh Rud catchment, gravity and pressurised irrigation methods were defined. In MIKE Basin, the parameterisation of irrigation methods can be conducted according to several parameters, specifying in what way and at what time a given field is irrigated through:

- The wetting faction—corresponding to the fraction of the field surface wetted during irrigation.
- The spray loss—the water fraction that evaporates before the water reaches the soil surface.
- The trigger—determines when the irrigation will start, whereby several trigger options are available in MIKE Basin:
- For the "FAO56 Irrigation Model", the option "Readily Available Water" (RAW) was chosen. The irrigation starts when the soil moisture content reaches the specified fraction of RAW, with RAW being defined as the volume of water that can be transpired by the crop without exposing the crop to soil water stress. The relationship between the amount of RAW and the "Total Available Water" (TAW) is given by:
- RAW = p × TAW, where p is the depletion fraction of the TAW at which soil moisture stress will start to reduce crop transpiration.

Table 16.1 Irrigation networks of the Zayandeh Rud catchment

Irrigation network	Total area (km^2)	MIKE Basin nodes
Abshar left	449.8	2
Abshar right	170.8	2
Borkhar	579.9	2
Chadegan	150.1	1
Faridan and Fereydunshahr	490.6	1
Karvan downstream Khamiran	14.7	1
Karvan upstream khamiran	89.8	1
Kordolia	34.5	1
Lenjanat downstream Lenj	74.3	2
Lenjanat upstream Lenj	60.9	2
Mahyar and Jarghoyeh	378.1	2
Nekouabad left	293.6	2
Nekouabad right	166.9	2
Roodasht North	282.6	2
Roodasht South	219.4	2

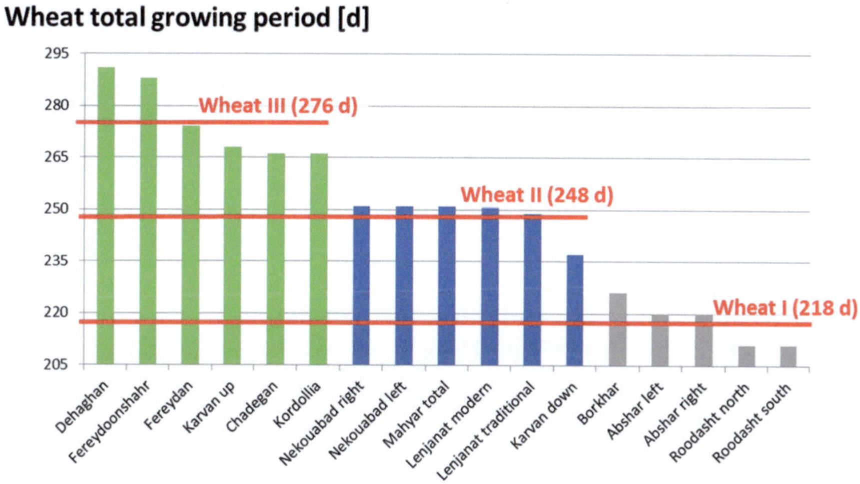

Fig. 16.3 Wheat generalisation and corresponding growing period

- The application option—calculates the application depth once the irrigation is initiated. Several options are available in MIKE Basin:
- For the "FAO56 Irrigation Model", the option RAW was chosen. The irrigation stops once the soil moisture content reaches the specified fraction of RAW.

The specific irrigation method has to be assigned to every crop and subsequently to every field with its specific area. For all crops and orchards, the gravity irrigation method has been used in the model, while additionally pressurised irrigation methods are used for some cultivation groups.

In general, for gravity irrigation it takes approximately six days to irrigate the entire network. Although this time may vary between the networks, this one unique value was applied for each network. For the pressurised irrigation, this feature of MIKE Basin is not applicable.

The soil characteristics are represented in MIKE Basin through the "soil water model", which quantifies the amount of water flowing between different layers in the soil profile and its water content storage, while tracking the amount of soil water available for soil evaporation and crop transpiration during the simulation. There are different types of soil water models available in MIKE Basin. In this model setup, the "FAO 56 soil water model" and the "Rice water and soil model" were used. For each of the soil types, the soil water model type "FAO 56 soil water model" was chosen and parameterised. This involves describing the soil moisture content (through the field capacity, wilting point and initial soil moisture content), the depth of the evaporable layer and the porosity. The data used is consistent with the soil characteristics applied in SWAT (Arnold et al. 1998). To increase the performance of the MIKE Basin model, only the soil type with the highest percentage was chosen for each irrigation scheme.

For climate and reference ET model selection, it was decided to use time series of rainfall and potential evapotranspiration produced by hydrological model SWAT. The rainfall and potential evaporation for each irrigation network was taken from a certain SWAT sub-basin. In order to choose the appropriate sub-basin, an analysis of the spatial variability of rainfall and potential evaporation was conducted (i.e. for summer period, see Fig. 16.4).

16.3 Water Demand Modelling for Municipal and Industrial Water Uses

Figure 16.5 provides an overview of the kind and location of different water uses in the Zayandeh Rud basin.

The municipal water demands have been included in the MIKE Basin model as regular water users:

- Water supply to Isfahan and Yazd
- Felman wells
- Tiran transfer
- Shahrekord supply

The Felman wells are known for their good water quality, providing municipal water supply for Isfahan as well partly for industry. It is assumed that 70% of the source of production water is from bank filtration. This has been applied to the time series of well output.

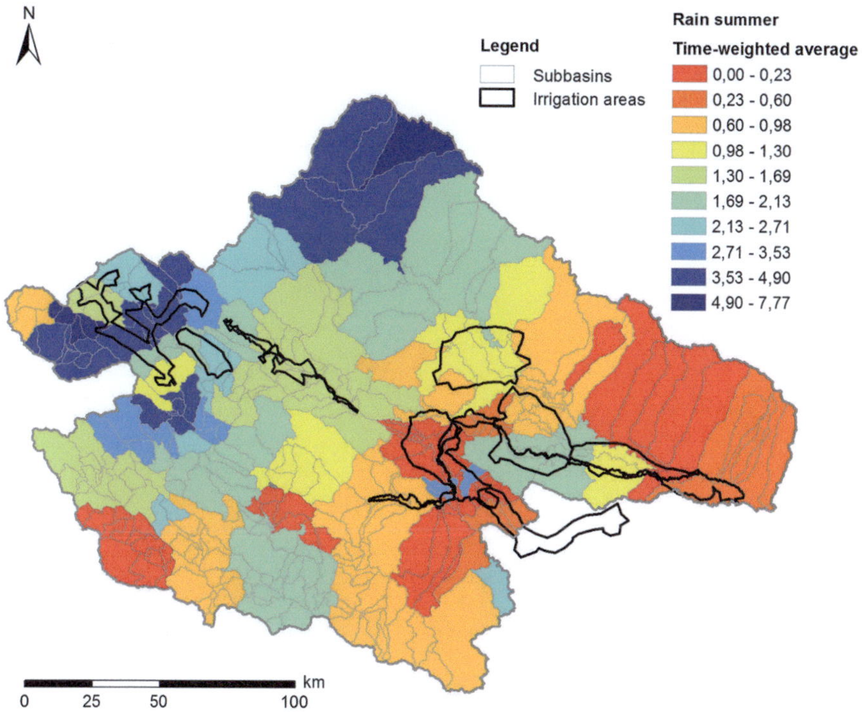

Fig. 16.4 Representative time-weighted rainfall distribution of the summer period

Waste water treatment plants (WWTPs) are part of the municipal water use. Most of them do not discharge directly into the Zayandeh Rud River, with the exception of the Baghbahadoran, Varzaneh, Zarin Shahr and Isfahan South WWTP. From all WWTPs, a selection was also made regarding those with a significant high nominal capacity. Therefore, only the Isfahan South WWTP was represented in the model, since it had an annual load of 57,552,000 m^3/a in 1390 as well as discharging directly into the Zayandeh Rud river, downstream of the Abshar distribution facility.

The current industrial water demand from surface and groundwater considered in the model can be seen in Table 16.2.

Existing industrial water users were excluded based upon the following considerations:

- users where the water abstraction was lesser than 10 l/s;
- abstractions that do not use water directly from the Zayandeh Rud river or from one of the main irrigation channels;
- groundwater abstraction too far away from the Zayandeh Rud river (no influence on the discharge); and
- supply is provided from urban water sources (which is already considered by demand of WTP Isfahan and must not be double accounted for).

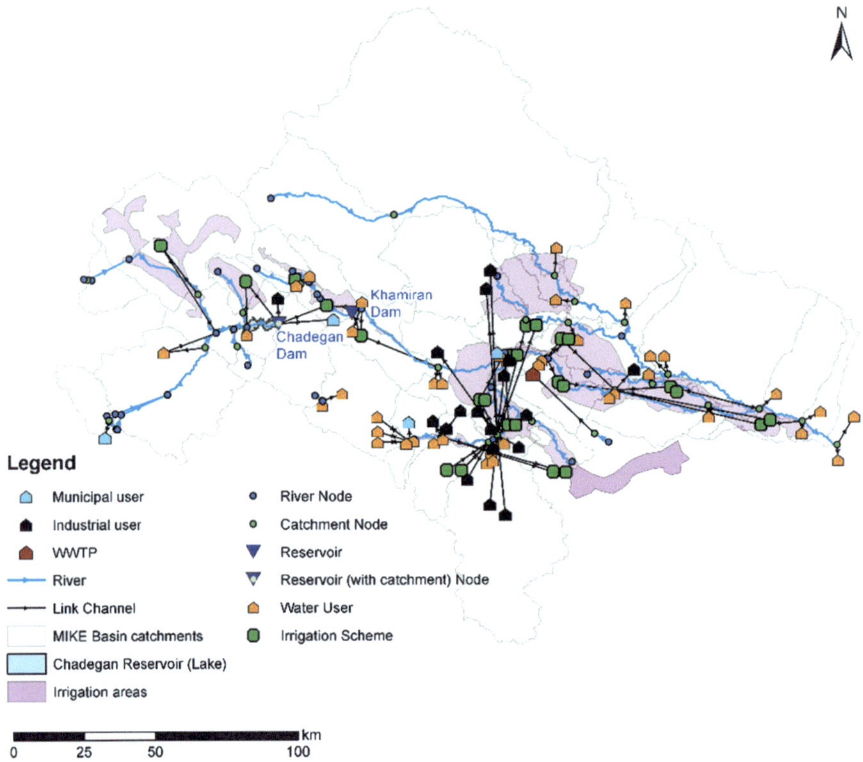

Fig. 16.5 Water users within the Zayandeh Rud catchment

The industrial water supply is assumed to have no returning water as a surplus from processing water. All surplus is used for gardens in the surroundings of the industry.

16.4 Interaction of Surface and Groundwater

Regarding agriculture, for each irrigation scheme the prioritisation of the water supply can be set up. Surface water always has the first priority, followed by groundwater. Accordingly, the groundwater extraction only occurs if there is a water deficit in surface water. The implementation of the groundwater in MIKE Basin is actually possible through linear reservoir parametrisations, whereby an aquifer can supply irrigation nodes. Using this MIKE Basin feature, the aquifer must comply to undisturbed natural interaction with surface water. This obviously is not the case in the Zayandeh Rud basin. Therefore, groundwater contribution was included through additional supply catchments.

Table 16.2 Industrial water users within the Zayandeh Rud catchment area, extraction from surface water (SW) corresponds to a normal and wet year

Name	Water source	Extraction SW [m^3/s]	Included in MB
7 Tir Complex	Zayanderoud River (BankFiltration)	0.03	Yes
Bafnaz Textile Co.	Borkhar Canal	0.02	Yes
Dolatabad	Urban water	0.00	No
Esfahan Steel Company	Zayanderoud River	0.82	Yes
Great area of Harand	Groundwater, Urban water, surface water	0.00	No
Great area of Jey	Groundwater, Urban water	0.00	No
Great Area of Najafabad	Nekooabad canal left, Urban water	0.02	Yes
Iran Telecommunication	Nekooabad Canal Left	0.08	Yes
Isfahan cement Co.	Nekooabad Canal Right	0.03	Yes
Isfahan Oil Refinery	Zayanderoud River, Urban water	0.21	No
Kamshjeh	Borkhar canal, Groundwater, Urban water	0.01	No
Mahmoodabad	Urban water	0.01	No
Mobarakeh Steel Company	Zayanderoud River	0.78	Yes
Other Industry 1	North RooDashtein Canel	0.21	Yes
Other Industry 2	Nekooabad Canel Left	0.15	Yes
Other Industry 4	Borkhar Canal	0.03	Yes
Petrochemical Co.	Borkhar- Shahinshahr/ Urban water	0.08	No
Power Plant Islamabad	Zayanderoud River, Urban water	0.38	Yes
Power Plant South	Zayanderoud River (BankFiltration)	0.04	Yes
Pulp and Paper Industry Co. Atrac	Zayanderoud River (BankFiltration)	0.04	Yes
Sepahan cement Co.	Zayanderoud River	0.03	Yes
Shahid Montazeri Plant	Borkhar Canal, Urban water	0.03	Yes
Synthetic Fibre raw materials	Zayanderoud River (BankFiltration)	0.06	Yes

Their contribution to the water supply of irrigation networks from groundwater is limited. The capacity value stands for the maximum amount of groundwater that can be extracted by the groundwater wells of a particular irrigation network. This maximum is related to the number of wells and their hydraulic capabilities. To obtain this maximum extractable amount of groundwater, a statistical analysis of time series of groundwater use for each irrigation network was conducted. It is assumed that the given time series contains the maximum extractable amount of groundwater. In order to avoid the use of outliers from measurement uncertainties, the 95% percentile of the time series was used as the maximum extractable amount of groundwater.

The seepage from the Zayandeh Rud River was considered in three separate sections and calculated in FEFLOW (Diersch 2014), using the following methodology:

- the budget rate was estimated for each river node to quantify the exchange in the river bed;
- the river length and width was estimated to calculate the river area; and
- the seepage quantity was calculated.

In the Zayandeh Rud basin, the degree of the above-described interaction of ground and surface water causes heavy cycle water. Therefore, the irrigation efficiency for a single field is much lower than for significantly large areas like the entire basin, as well as study units or irrigation networks. For the latter, it can be assumed to be almost 1.

Accordingly, the irrigation demand as a whole for such a large area should be calculated as net irrigation demand. This demand is much lower than the total water use of all fields. The algorithm for MIKE Basin calculates the irrigation demand for each network based upon the demand of the plant, the deficit in the ground and the losses on the field surface due to evaporation. Therefore, the above-derived requirement is fulfilled.

16.5 Dam Management and Water Supply Rules

There are two major dams within the Zayandeh Rud catchment, namely the Zayandeh Rud Dam—forming the Chadegan reservoir—and the Khamiran Dam. Both have been included in the MIKE Basin model and their main parameters are shown below in Table 16.3.

For water availability in the Zayandeh Rud dam, water transfers from neighbouring Karun basin are important, with the following being considered:

- transfer from Cheshme Langan;
- Koohrang tunnels 1 and 2.

Table 16.3 Reservoir properties assigned for the reservoirs in MIKE Basin

[m.a.s.l.]	Zayandeh Rud Dam	Khamiran Dam
Initial water level	2052	2015
Bottom level	1000	1000
Top of dead storage	1983	1996
Dam crest level	2063	2018
Flood control level	2063	2018

Table 16.4 Area of water rights within irrigation networks

Irrigation network	Percentage of water right of total cultivation area [%]		
	Haghabe	Sahmabe	Hagh eshteraki
Anhar sonati downstr. Pol Zamankhan, up & down Lenj	100	0	0
Mahyar & Jarghoyeh	0	0	100
Nekouabad left	30	70	0
Nekouabad right	30	70	0
Borkhar	0	0	100
Abshar left	45	48	7
Abshar right	45	48	7
Roodasht North	80	0	20
Roodasht South	80	0	20

The non-standard water management rules for the Zayandeh Rud Dam and the water supply rules according to the prioritisation and water rights framework are a matter of customisation of MIKE Basin. The parameters of the water management rules are reviewed and updated at the beginning of summer and winter.

The key elements are:

- calculating the required amount of water for four supply levels with increasing degree of demand fulfilment; and
- establishing the highest admissible supply level based upon the acceptable storage change in the Zayandeh Rud dam.

The choice of the highest acceptable supply level is based upon the admissible storage reduction in the summer and autumn, just as the required storage increases in winter and spring time. From the analysis of the historical storage change analysis during these seasons, the basis values for the storage change were established. Their values are adopted at the beginning of the summer or winter by a scaling factor, reflecting the mismatch between the expected and realised storage change in the previous season. The degree of this scaling can be controlled through a critical scale coefficient if a value falls below the critical storage content.

The water supply of each irrigation network is realised depending on the supply priority of irrigated fields and their area. Therefore, for each irrigation network it was determined how much area belongs to which class of water right (see below Table 16.4).

In Table 16.4, the name of each water right class is shown in Farsi. Haghabe requires highest priority of water supply, while Hagh eshteraki requires the lowest. A priority of water supply for each irrigation network results from the area of these water right classes.

16.6 Plausibilisation Setup

The above-described model setup regarding data and rules resulted in MIKE Basin Zayandehroud as shown below in Fig. 16.6. It should be noted that in this MIKE Basin model all dates are according to the Persian calendar. Since years earlier than 1900 cannot be shown in Microsoft Office and MIKE Basin, the value 1000 was added to each Persian calendar year. Thus, the Persian year 1387 (2008) is converted into 2387 and any display of the year 2387 corresponds to the Persian year 1387.

Based upon the data from 1379 (2000) to 1387 (2008), the results for the storage content, the release from the dam and the supply for one of the most relevant irrigation systems are shown in Figs. 16.7, 16.8, and 16.9.

It is evident from the modelling results that it is almost impossible to represent the storage management with only one parameter set for the period from September 2000 to August 2008. Furthermore, there are significant uncertainties in the measured data since a part of the usage values are only available as yearly amounts. In order to perform monthly simulations, the yearly values were disaggregated to monthly values based upon various assumptions.

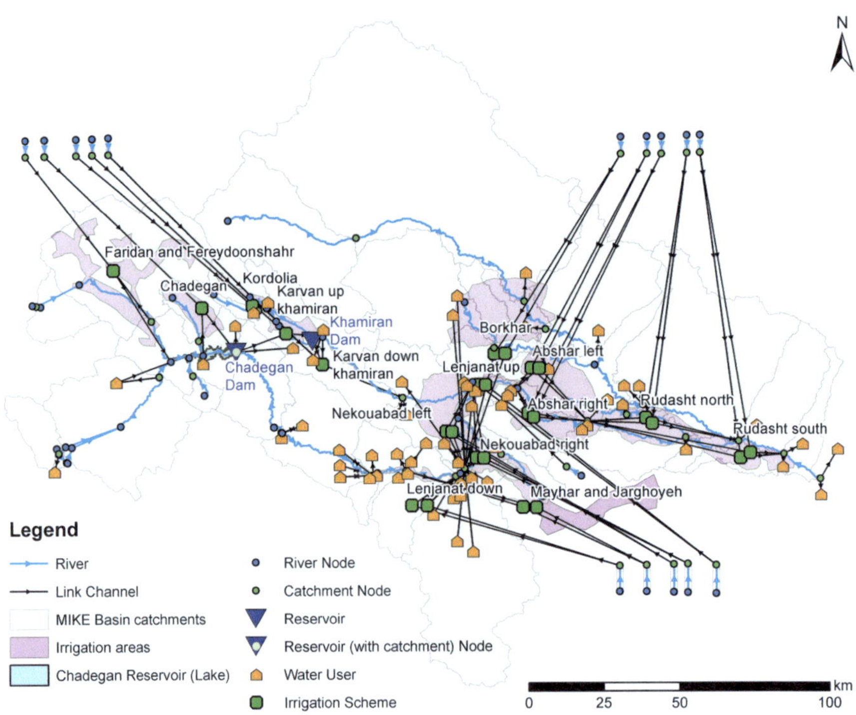

Fig. 16.6 MIKE Basin model setup

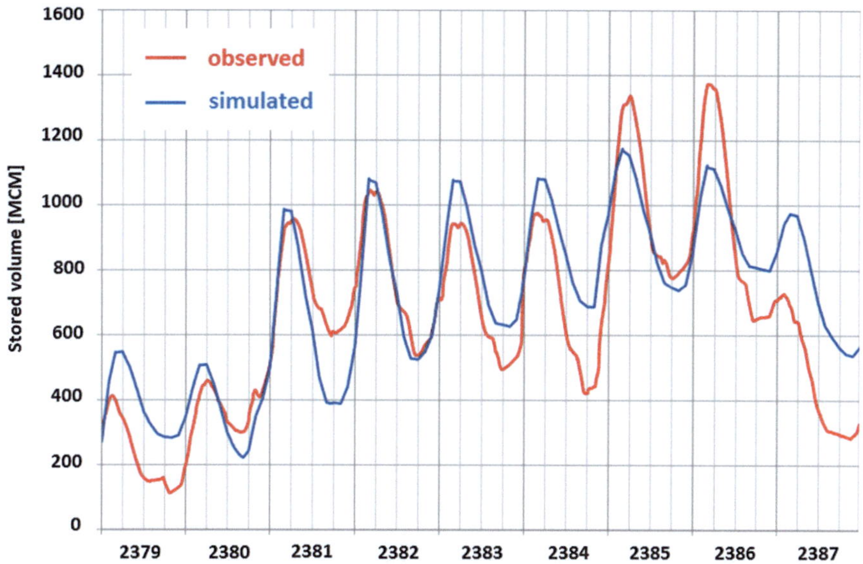

Fig. 16.7 Observed and simulated storage content of the Zayandeh Rud Dam

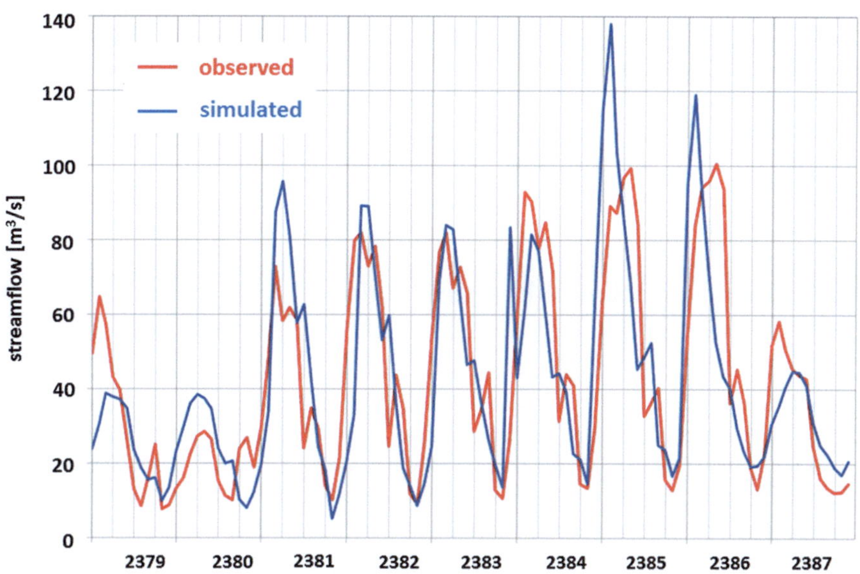

Fig. 16.8 Observed and simulated run-off of the Sade Tanzimi

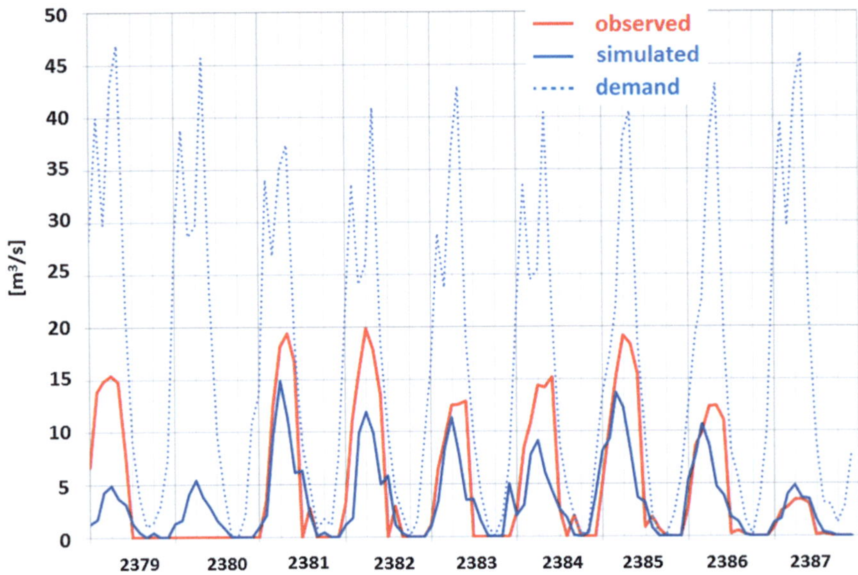

Fig. 16.9 Observed and simulated run-off and the actual demand at the Nekouabad left irrigation system

It should be noted that the quality of water management models can only be adjusted beyond the calibration of the run-off model by changing water use data or management and supply rules. A calibration of the latter is neither sensible nor possible, since this would inadmissibly change the measured or estimated values.

16.7 Demonstration of Practical Value

The degree of water stress in the Zayandeh Rud basin may be demonstrated by the impact on surface water availability if a prioritised compliance of a 5 m^3/s discharge to Abshar (located at the downstream end of Isfahan) were required.

In Figs. 16.10 and 16.11, the storage content of the Zayandeh Rud Dam and the discharge of the gauging station Pol Choum (located downstream of Abshar) for the current state and the case called "scenario 1" are compared.

The results show that with the current demand requirements—particularly from agriculture—the minimum flow of 5 m^3/s over-stresses the systems water management. Accordingly, within seven years the storage content is completely emptied twice.

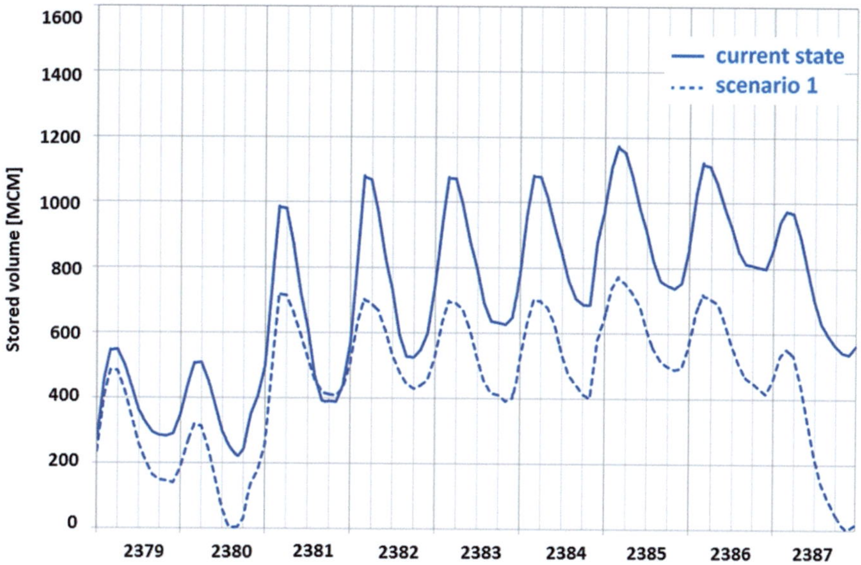

Fig. 16.10 Comparison of the storage content of the Zayandeh Rud Dam for the current state and for scenario 1

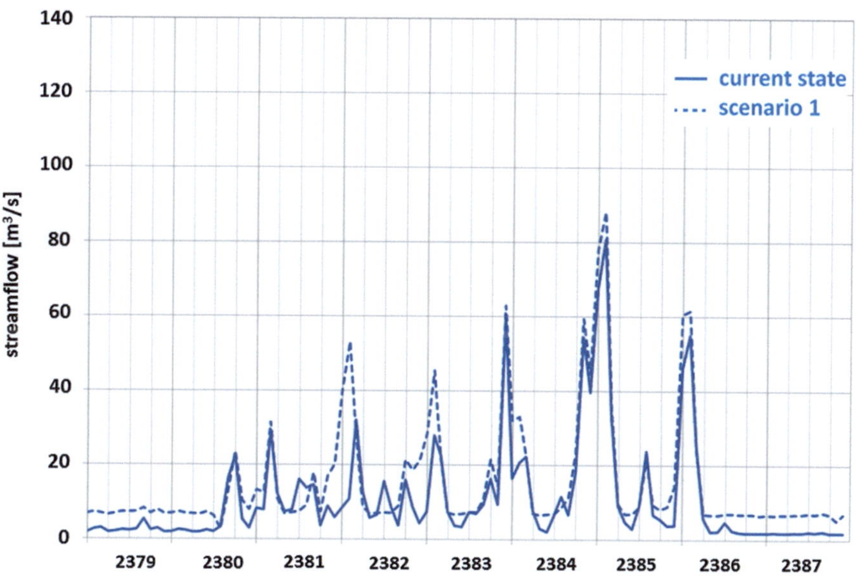

Fig. 16.11 Run-off at the gauging station Pol Choum for the current state and for scenario 1

References

Allen RG, Pereira LS, Raes D, Smith M (1998) Crop evapotranspiration: guidelines for computing crop water requirements. FAO Irrigation and Drainage Paper No 56, FAO, Rome

Arnold JG, Srinivasan R, Muttiah RS, Williams JR (1998) Large area hydrologic modeling and assessment. Part 1: Model development. J Am Water Resour Assoc 34(1):73–89

DHI (2012) MIKE Basin—a modelling system for river basin management and planning. Model documentation—MIKE by DHI 2012

Diersch H-J (2014) FEFLOW. Finite element modeling of flow, mass and heat transport in porous and fractured media. Springer, Berlin. doi:10.1007/978-3-642-38739-5